Lecture Notes in Mobility

The book series Lecture Notes in Mobility (LNMOB) reports on innovative, peer-reviewed research and developments in intelligent, connected and sustainable transportation systems of the future. It covers technological advances, research, developments and applications, as well as business models, management systems and policy implementation relating to: zero-emission, electric and energy-efficient vehicles; alternative and optimized powertrains; vehicle automation and cooperation; clean, user-centric and on-demand transport systems; shared mobility services and intermodal hubs; energy, data and communication infrastructure for transportation; and micromobility and soft urban modes, among other topics. The series gives a special emphasis to sustainable, seamless and inclusive transformation strategies and covers both traditional and any new transportation modes for passengers and goods. Cutting-edge findings from public research funding programs in Europe, America and Asia do represent an important source of content for this series. PhD thesis of exceptional value may also be considered for publication. Supervised by a scientific advisory board of world-leading scholars and professionals, the Lecture Notes in Mobility are intended to offer an authoritative and comprehensive source of information on the latest transportation technology and mobility trends to an audience of researchers, practitioners, policymakers, and advanced-level students, and a multidisciplinary platform fostering the exchange of ideas and collaboration between the different groups.

Peter Moertl · Bernhard Brandstaetter
Editors

Shaping Automated Driving to Achieve Societal Mobility Needs

A Human-Systems Integration Approach

Springer

Editors
Peter Moertl
Virtual Vehicle Research GmbH
Graz, Austria

Bernhard Brandstaetter
Virtual Vehicle Research GmbH
Graz, Austria

ISSN 2196-5544 ISSN 2196-5552 (electronic)
Lecture Notes in Mobility
ISBN 978-3-031-52552-0 ISBN 978-3-031-52550-6 (eBook)
https://doi.org/10.1007/978-3-031-52550-6

This Springer imprint is published by the registered company Springer Nature Switzerland AG
The registered company address is: Gewerbestrasse 11, 6330 Cham, Switzerland

Paper in this product is recyclable.

Contents

Shaping Automated Driving to Meet Societal Mobility Needs: The HADRIAN Project

Peter Moertl and Bernhard Brandstaetter

Abstract Research and development of high levels of automated driving (AD) vehicles has received considerable attention in recent years. Thereby, for the time being, humans will remain actively involved to assure overall safety, whether as drivers, safety drivers, or tele-operators. This effectively shifts human tasks and responsibilities compared to manual driving. To make these shifts as safe and comfortable as possible and also reliable and predictable to use, the Horizon Europe research and innovation project HADRIAN (Holistic Approach for DRiver role IntegrAtioN) investigated and evaluated holistic, user-centered solutions. Thereby, the HADRIAN consortium envisioned a larger eco-structure from which it would be possible to reconceptualize what is part of driving automation and how it works. This meant to include parts of the roadside information infrastructure as well as to directly include the human driver in two specific ways: first by designing AD solutions that directly support their mobility needs and constraints. And secondly, by shaping the AD solutions that allow drivers to perform their new tasks and responsibilities more safely and comfortably. In this volume we describe how such holistic, user centered approach allows to derive better and more powerful solutions than those that are merely focused around the vehicle. For this we report the results of a series of innovations and their evaluations and demonstrations in the field. We conclude with how such approach also requires more tightly connected and inter-disciplinary team collaborations than are often found in current research and development organizations. In this first chapter we introduce the underlying human factors problems of currently available levels of AD and thereby motivate the starting point for the holistic user-centered approach and solutions that are then described to greater extent in the following chapters.

P. Moertl (✉) · B. Brandstaetter
Virtual Vehicle Research GmbH, Graz, Austria
e-mail: Peter.Moertl@v2c2.at

P. Moertl and B. Brandstaetter (eds.), *Shaping Automated Driving to Achieve Societal Mobility Needs*, Lecture Notes in Mobility,
https://doi.org/10.1007/978-3-031-52550-6_1

1 Introduction

Highly automated driving promises to offer improved safety together with a multitude of previously unimagined possibilities such as reduced stress and more meaningful or productive activities while driving. Also, automated driving could enable mobility of otherwise excluded mobility participants such as elderlies or people with disabilities. To make automated driving a reality in Europe, the public–private partnership CCAM (Connected Cooperative Automated Mobility) was formed in 2021. Thereby, CCAM has formulated the goal to develop user-centered mobility solutions that enable all-inclusive mobility, while also increasing safety, reducing congestion, and contributing to decarbonization (CCAM 2022). While these goals are clearly defined, their realization is experiencing delays and no wide-spread market penetration of higher levels automated driving vehicles is in graspable reach. A specific problem thereby relates to the new roles and tasks of the human driver to use, manage, and supervise current designs of AD vehicles..

1.1 Challenges of Current Driving Automation from the User Perspective

While there are worldwide examples for highly automated vehicles offering mobility service such as Waymo (https://waymo.com/), Pony.AI (https://pony.ai/), or Cruise (https://getcruise.com/rides/), currently a much larger field of developments focus on advanced driving assistance systems (ADAS) as an add-on to manually driven passenger vehicles such that drivers can choose between manual driving and automated driving at different levels. What are the challenges of such automation from the human driver and user perspective?

New vehicles of today are already often equipped with partial automated driving functions (SAE Level 2, SAE International 2021). At this automation level the automated driving (AD) vehicle takes over lateral and longitudinal maneuvers while the driver remains responsible to continuously monitor whether interventions or corrections are needed. Such monitoring tasks can be difficult to perform by humans as has been well established in research (e.g. Kaber and Endsley 1997; Rudin-Brown and Parker 2004; Stanton 2019; Stanton and Young 2005). Monitoring requires attention which can be effortful when maintained over extended periods of time. Thereby, the longer the AD vehicle appears to work successfully, the harder it can become for humans to stay engaged and remain sufficiently alert to detect automation or traffic problems and to appropriately intervene. This is referred to as automation conundrum (Endsley 2019): the better the automation works and the more often operators can rely on the automation, the more difficult it gets for them to recover when automation failures occur. Therefore, in the end, drivers get bored, engage in other tasks, or create unsafe shortcuts. This can decrease safety.

The next higher level of automated driving after SAE L2 is conditional automated driving (SAE L3, SAE International 2021). Here, the driver can disengage from driving for some periods and the AD vehicle takes over lateral and longitudinal control and, in addition, monitors the environment for situations that require control by the human driver: these are situations when the AD vehicles reaches the end of its operational design domain (ODD) (SAE International 2021). During periods of SAE L3 driving, the driver becomes a user, free to perform non-driving related activities (NDRA) such as reading or watching a movie. However, eventually the user will have to become a driver again and is responsible for taking back control from the AD vehicle where the driver is responsible for remaining ready for such fallbacks. Here the challenge for the driver consists of getting back into the safety critical control loop of driving which can be difficult for any operator who has been disengaged for a while. This is also known as the out-of-the-loop problem and has been identified as difficult and error prone in several domains such as in aviation (see e.g. Endsley 2016; Hancock 2019; Kaber and Endsley 1997). Furthermore, it is currently not possible for an AD vehicle to guarantee the driver a certain time period to take back control: the amount of time for the vehicle sensors to detect an unexpected obstacle or event depend highly on the visibility in the surrounding environment. Therefore, the human driver is challenged to flexibly adapt to a variety of potentially quite surprising situations. This again can decrease safety.

Also, both automated driving levels, SAE L2 and SAE L3 may become available on the same trip. For example, a driver may have to drive manually first to bring the vehicle into a geographic area that meets the ODD for automated driving. At this point, the driver may engage automated driving, then disengage it again, engage another AD level and so forth. This spells trouble because these two automated driving levels require different tasks and responsibilities of the driver and therefore require the driver's appropriate automation mode awareness. This can be a problem because differentiating multiple modes of automation can be difficult for any operator. Even highly trained aircraft pilots sometimes get confused about the mode the automation is in. When Asiana flight 214 attempted to land in San Francisco Airport in 2013, the three pilots who were in the cockpit at the time missed to detect that the airplane was in an automation mode intended for enroute operation at high altitude rather than in the mode they wanted the airplane to be in when landing (NTSB 2013). The plane crashed on the runway as result of these three pilots lacking sufficient automation mode awareness of the aircraft they were flying. Drivers of passenger vehicles are generally much less well trained than airline pilots and also generally less thoroughly certified. Therefore, the potential of error is even higher, this again can decrease safety.

So in summary, in current AD vehicles we see human-systems integration challenges to appropriately accommodate the new tasks and responsibilities of the human driver to be safe and comfortable. This challenge holds us back from reaching the promised goals of AD vehicles on European road and gave rise to the EU project HADRIAN.

1.2 The EU Project HADRIAN

The above described human-systems integration challenges led the European Union to publish a call for research as part of the H2020 program to identify possible solutions: how can we design AD vehicle systems that allow everyday drivers to safely use and differentiate between the different automation modes? What solutions can increase the safety and comfort of AD vehicles from a user perspective?

The HADRIAN consortium (https://hadrianproject.eu/) applied in 2019 for this research call and proposed a (1) holistic as well as (2) user-centered approach to solve the described problems.

(1) The starting assumption for the HADRIAN approach was to envision a new and larger eco-system from which it would become possible to conceive user-centered solutions that go beyond the AD vehicle itself. Such holistic system seemed needed because the described human-systems integration challenges seemed a direct result of focusing AD developments mainly on the vehicle per se, see Fig. 1. The upper graph in Fig. 1 depicts the status quo for manually driven vehicles where the human driver maneuvers the vehicle for safe operations on the road and traffic environment. For automated vehicles as of today, the maneuvering of the vehicle is automated (lower left graph in Fig. 1), but the other tasks remain the responsibility of the human (i.e. monitoring during SAE L2 and take-over for SAE L3, as outlined above). Thereby, human drivers in AD vehicles of today retain similar responsibilities as in manually driven vehicles while some of their tasks change significantly. This represents a significant challenge to the acceptance and safe use of AD vehicles that is not easily solved through innovations of the AD vehicle per se. Therefore, the HADRIAN consortium took a step back and searched for solutions in a larger space to find more comprehensive solutions to address the misalignment of human tasks and responsibilities, see right-hand graph in Fig. 1: the holistic Automation Systems (DAS).

(2) The second main assumption of the HADRIAN approach is that creating a holistic, holistic more powerful DAS requires also shaping it in ways that users will find attractive and meet their mobility needs. For what purposes would users want to use AD vehicles. While user-centered design is commonly applied when developing software-based systems, we apply this approach to the development of AD vehicle functions. As will be described in Chap. 2, the HADRIAN project identified user-needs and constraints to formulate use scenarios and personas that led to define user requirements. These user requirements formed the basis to design and iteratively refine solutions within the larger, holistic DAS.

The HADRIAN consortium received funding from the EU and developed and evaluated the holistic, user-centered approach for DAS between Dec 2019 and May 2023. This book summarizes the project approach and the most important lessons learned from the project. Specifically, the holistic DAS is described in the following subsection, the user-centered design approach is described in the next chapter.

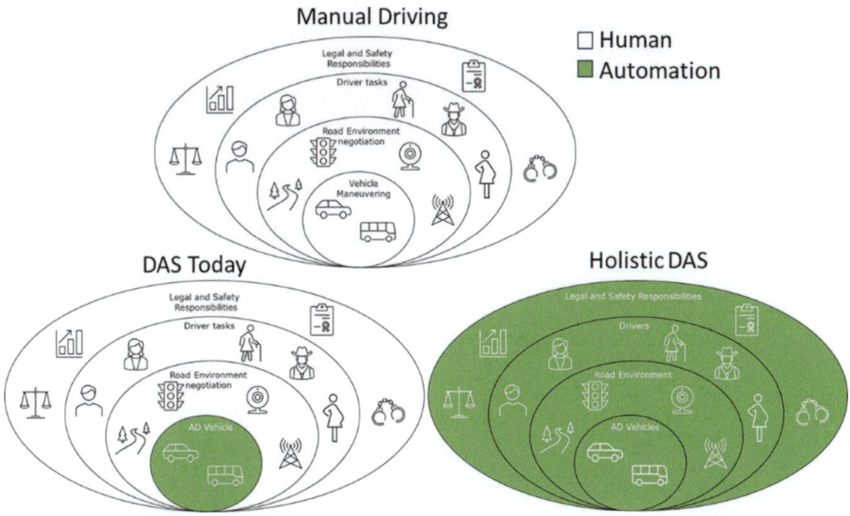

Fig. 1 Expanding the system scope from manual driving toward holistic driving automation systems (DAS)

2 From AD Vehicles to Holistic DAS

A holistic DAS in the HADRIAN project includes the road information infrastructure, the vehicle, and the driver. This is shown in Fig. 2 where the proposed enhancements of the DAS are explicitly depicted:

2.1 *Expanding the DAS to Include Road Information Infrastructure*

The operational design domain (ODD) for an AD vehicle consists of the operating conditions for its functioning (SAE International 2021). This can include the type of road such as a motorway, the absence of rain, snow, or obstructions, as well as intact lane markings. When the vehicle sensors sense conditions that are not consistent with the AD vehicle's ODD for SAE L3, the vehicle requests the driver to take back manual driving. This may surprise the driver and could lead to rushed, unprepared responses with potentially unsafe outcomes. Also, having to perform sudden time-critical safety maneuvers may also limit the experience of comfort and well-being of the driver. Such surprises could be reduced if road information infrastructure predictively informed the AD vehicle about upcoming ODD changes so that the vehicle could thereby guarantee drivers a minimum amount of time to transition back to manual driving. This should not only allow drivers to take back driving controls in a

Fig. 2 The Holistic HADRIAN DAS includes the road information infrastructure, the vehicle, and the driver

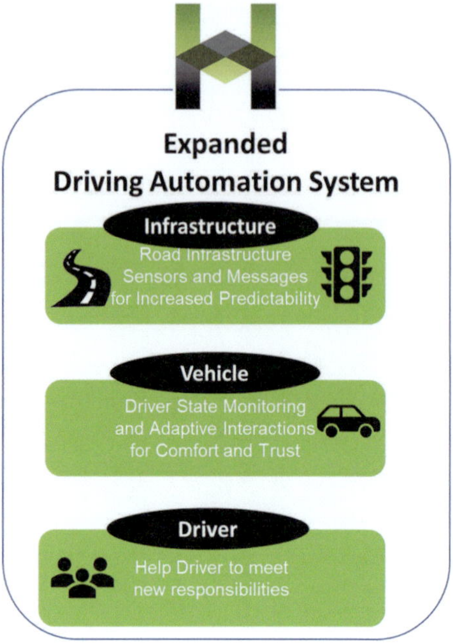

safe way but also learn to better and more consistently perform the appropriate take-over maneuvers. The integration of road infrastructure information into the DAS forms the first pillar of the HADRIAN holistic DAS.

2.2 *Expanding the DAS Toward Adaptive, Fluid In-Vehicle Interactions*

The second pillar of the HADRIAN holistic DAS consists of fluid and adaptive human–computer interactions (HCI) to help drivers understand and use the AD vehicle and increase their comfort. In today's manually driven vehicles the driver performs all the adaptations to use and control the vehicle dependent on the traffic and environmental situation (see left in Fig. 3). Beside the actual planning, navigation, and control of the vehicle, the driver has to adjust the various vehicle settings and options (see e.g. Bubb et al. 2015; Michon 1985). As modern vehicle systems are becoming increasingly complex as result of automation, drivers need to know more and more about the automation to effectively use the vehicle. Therefore, the driver's task load for performing these adaptations increases and modern vehicles have already many features and options that often remain unused.[1] Remember renting the last vehicle and the effort of figuring out how to open the fuel cap or engaging

[1] https://www.motorbiscuit.com/drivers-dont-use-most-tech-features-new-cars/

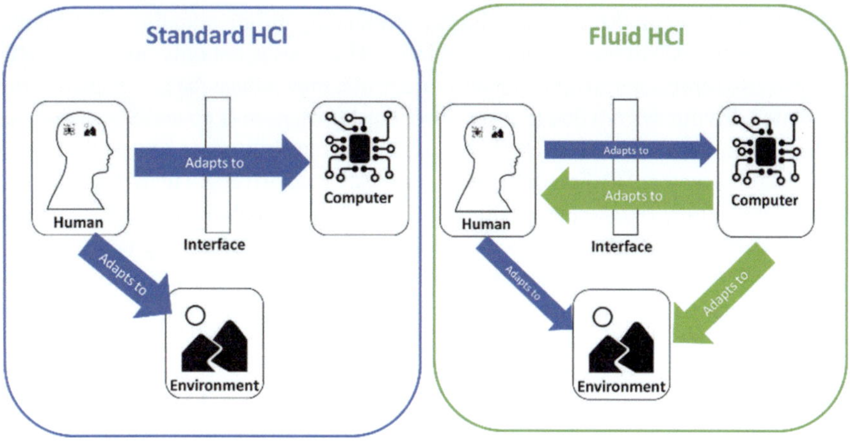

Fig. 3 Fluid HCI reduce the human adaptation burden by adapting to the Human

the hazard lights? It takes time and effort for humans to learn to use these vehicle features, even more so when they are hidden below several layers of digital menus. Also older drivers may have additional difficulties to adapt to novel technologies, as will be further described in Chap. 2.

The fluid in-vehicle interactions are intended to reduce the human load for many of these adaptations and moves these adaptation to the vehicle automation (Dijksterhuis et al. 2012; Pretto et al. 2020). Thereby, the vehicle dynamically adapts to the driver state and needs of the user as well as to the situation (see left side in Fig. 3). Instead of burdening the driver with cognitive and motoric adaptation tasks, the vehicle performs these adaptations. However, such adaptations could be confusing if they contradict the driver's expectations and need to be made appropriately transparent to the user (Feigh et al. 2012). Therefore, they have to be carefully designed, planned and tested, requiring a level of knowledge about the driver's background and competences that goes beyond what is needed by designers of static HCI's that do not adapt to the state of the users.

One prerequisite of fluid interactions is for the vehicle to know enough about the driver's state and the current situation. This requires sufficient in-cabin sensors that observe the driver and external sensors with information about the road and traffic environment. The quality and reliability with which the driver state can be identified represents a fundamental boundary for fluid interactions.

2.3 *Expanding the DAS to Increase Driver Competences*

The third pillar of the HADRIAN approach helps human drivers to safely engage, use and disengage the DAS. In today's AD vehicles, drivers often learn to use the automation features of their vehicles through "trial-by-error" strategies (Ebinger et al.

2023; Neuhuber et al. 2022). This entails experiencing unsafe situations that exhibit the limits of the AD to learn about these limits. This causes not only potential safety risks but also represents an unfortunate strategy that may reduce the acceptance of the automated driving system due to insufficient trust formation (Lee and See 2004). As drivers first overtrust the AD vehicle in terms of its capabilities and then experience repeated unexpected AD disengagements, this may lead drivers to undertrust the AD vehicle which may reduce their willingness to use the automated driving functions. Therefore, human-centered technology developments aim at forming appropriately "calibrated" trust (Hoff and Bashir 2015) as fundament for sustained product use. Also, a driver who operates the AD vehicle functions incorrectly ("misuse" the vehicle) increases safety risks. In the design of current AD vehicles, manufacturers have no way to address such possible misuse.

To address these current limitations, the third pillar of the HADRIAN approach actively increases the knowledge and competences of drivers to safely and reliably use the AD vehicle functions and also to actively check whether the driver uses them appropriately. In addition, if safety–critical inappropriate use is detected it helps drivers to avoid such behavior. This is achieved through a real-time, digital, in-vehicle tutoring application that informs the driver through visual and auditory means about how the AD system works, how it can be safely engaged, used, and disengaged, and how the driver can improve performance, if needed. Especially if, during the transition from conditional automated driving to manual driving the driver is detected to take back control in a haste without sufficiently checking the environment, the driver receives an auditory recommendation on how to improve the take-over performance next time.

3 Outlook

The HADRIAN project postulates a holistic, user-centered approach to shape AD vehicles according to the specific mobility needs of users and make them safe and comfortable to use. In the remainder of this book we will expore this approach in detail but is it realistic to build such an eco-system? The current paths of evolving vehicles has been well established over the years and it seems to require a huge coordinated effort to create such a large holistic system that includes humans, vehicles, and humans. However, there are many examples where such large eco-systems have been developed over time such as the internet, international air travel, banking, mail delivery as well as railroad, and aviation systems. In the early age of aviation, for example, did pilots fly and land airplanes purely based on visual orientation, using their eyes to orient themselves. Soon, however, ground based methods were developed that communicated position and navigation information to the airplane, (e.g. via radio beacons), allowing pilots and later automation, to continue flying and landing even under low visibility conditions. In other words, the plane's ODD was considerably enhanced to meet customer needs for air travel by expanding the system boundaries of the airplane. This expanded air travel beyond previous imagination and provided benefits that in the end paid for the large costs of this extensions that consists of airport infrastructure, radars, beacon systems, air traffic control, etc. Similar could be applied to the development of automated vehicles. However, in order to decide whether it is worth creating such larger eco-system, we need to know about its possible benefits. This is what the HADRIAN project intended to do: to investigate and quantify some of the possible benefits but also to test and demonstrate its feasibility. We report this in the following chapters.

4 Overview of the Following Chapters

Chapter two describes the human-centered design process, the other critical pillar of the HADRIAN project. After a review of European mobility visions about the expectable path of AD vehicles within the larger landscape of multi-modal mobility developments, three exemplary mobility personas were identified that are expected to benefit the most from AD vehicles based on current mobility plans in the EU: an elderly driver, a truck driver, and an office worker who extends work into the AD vehicle. For these personas, the consortium then derived a set of solutions to increase the quality of the DAS. These are here referred to as HADRIAN solutions.

In chapter three, a set of solutions for fluid, adaptive driver interactions is presented as part of an evaluation study in a driving simulator. The results of this driving simulator study provided insights into the final version of the HADRIAN solutions that were subsequently evaluated for a field-demonstration study.

In chapter four, a set of collaborative solutions that help drivers during manual driving to manage critical safety situations was evaluated in a driving simulator study that focused on enhancing the mobility of elderly drivers. Here, again, the studies provided important insights for finalizing the HADRIAN solutions for a field-demonstration study.

Chapter five describes the planning and results of a final field-demonstration study where the combined HADRIAN innovations were installed and demonstrated in two real vehicles. Also, the final conclusions are presented.

Acknowledgements The planning and execution of the HADRIAN project were a joint effort of many partners who all needed to work together and cross boundaries to achieve the project results. Therefore we thank all partners in the consortium for their contributions to realize this project. Specifically we want to thank Manuela Klocker from Virtual Vehicle who was key to successful project execution. Also, we thank Georgios Sarros as responsible project officer from CINEA for his thorough comments during reviews and supportive guidance. HADRIAN has received funding from the European Union's Horizon 2020 research and innovation programme under grant agreement No 875597. This document reflects only the author's view, the Climate Innovation and Networks Executive Agency (CINEA) is not responsible for any use that may be made of the information it contains. The publication was written at Virtual Vehicle Research GmbH in Graz and partially funded within the COMET K2 Competence Centers for Excellent Technologies from the Austrian Federal Ministry for Climate Action (BMK), the Austrian Federal Ministry for Labour and Economy

(BMAW), the Province of Styria (Dept. 12) and the Styrian Business Promotion Agency (SFG). The Austrian Research Promotion Agency (FFG) has been authorised for the programme management.

References

Bubb H, Bengler K, Grünen RE, Vollrath M (2015) Automobilergonomie. Springer Fachmedien Wiesbaden. https://doi.org/10.1007/978-3-8348-2297-0

CCAM (2022) CCAM strategic research and innovation agenda 2021–2027: European leadership in safe and sustainable road transport through automation. https://www.ccam.eu/wp-content/uploads/2022/05/CCAM_SRIA-report_web.pdf

Dijksterhuis C, Stuiver A, Mulder B, Brookhuis KA, de Waard D (2012) An adaptive driver support system: user experiences and driving performance in a simulator. Hum Fact J Hum Fact Ergon Soc 54(5):772–785. https://doi.org/10.1177/0018720811430502

Ebinger N, Trösterer S, Neuhuber N, Mörtl P (2023) Conceptualisation and evaluation of adaptive driver tutoring for conditional driving automation. In: Proceedings of the human factors and ergonomics society europe chapter 2023 annual conference. http://hfes-europe.org

Endsley MR (2016) From here to autonomy: lessons learned from human–automation research. Hum Fact. 0018720816681350

Endsley MR (2019) Situation awareness in future autonomous vehicles: beware of the unexpected. In: Bagnara S, Tartaglia R, Albolino S, Alexander T, Fujita Y (eds) Proceedings of the 20th congress of the international ergonomics association (IEA 2018), edn 824. Springer International Publishing, pp 303–309. https://doi.org/10.1007/978-3-319-96071-5_32

Feigh KM, Dorneich MC, Hayes CC (2012) Toward a characterization of adaptive systems: a framework for researchers and system designers. Hum Fact 54(6):1008–1024. https://doi.org/10.1177/0018720812443983

Hancock PA (2019) Some pitfalls in the promises of automated and autonomous vehicles. Ergonomics 1–17. https://doi.org/10.1080/00140139.2018.1498136

Hoff KA, Bashir M (2015) Trust in automation integrating empirical evidence on factors that influence trust. Hum Fact J Hum Fact Ergon Soc 57(3):407–434

Kaber DB, Endsley MR (1997) Out-of-the-loop performance problems and the use of intermediate levels of automation for improved control system functioning and safety. Process Saf Prog 16(3):126–131. https://doi.org/10.1002/prs.680160304

Lee JD, See KA (2004) Trust in automation: designing for appropriate reliance. Hum Fact J Hum Fact Ergon Soc 46(1):50–80

Michon JA (1985) A critical view of driver behavior models: what do we know, what should we do. Hum Behav Traffic Saf 485–520

Neuhuber N, Ebinger N, Pretto P, Kubicek B (2022) How am i supposed to know? Conceptualization and first evaluation of a driver tutoring system for automated driving. In: Proceedings of the human factors and ergonomics society europe chapter 2022 annual conference

NTSB (2013) Descent below visual glidepath and impact with seawall asiana airlines flight 214 boeing 777–200ER, HL7742 San Francisco, California July 6, 2013 (Accident Report NTSB/AAR-14/01). https://www.ntsb.gov/investigations/accidentreports/reports/aar1401.pdf

Pretto P, Mörtl P, Neuhuber N (2020) Fluid interface concept for automated driving. In: Krömker H (eds) HCI in mobility, transport, and automotive systems. Automated driving and in-vehicle experience design, edn 12212. Springer International Publishing, pp 114–130. https://doi.org/10.1007/978-3-030-50523-3_9

Rudin-Brown CM, Parker HA (2004) Behavioural adaptation to adaptive cruise control (ACC): implications for preventive strategies. Transp Res F: Traffic Psychol Behav 7(2):59–76. https://doi.org/10.1016/j.trf.2004.02.001

SAE International (2021) Taxomony and definitions for terms related to driving automation systems for on-road motor vehicles: (J3016)
Stanton NA, Young MS (2005) Driver behaviour with adaptive cruise control. Ergonomics 48(10):1294–1313. https://doi.org/10.1080/00140130500252990
Stanton NA (2019) Thematic issue: Driving automation and autonomy. Theor Issues Ergon Sci 1–7. https://doi.org/10.1080/1463922X.2018.1541112

User-Centered Design of Automated Driving to Meet European Mobility Needs

Carolin Zachäus, Sandra Trösterer, Cyril Marx, and Peter Moertl

Abstract Automated Driving (AD) technologies are transforming the mobility sector, promising enhancements in efficiency and safety as well as user experience tailored to diverse user needs. As the sector moves toward a more automated ecosystem, it's essential to recognize the varied expectations and requirements of its users. Elderly drivers, for instance, have distinct needs, from functional transportation to the aesthetic pleasure of driving. As age-associated impairments affect their driving capabilities, Advanced Driving Assistance Systems (ADAS) present an opportunity to augment their safety and confidence. Truck drivers or business people on the other hand have different sets of needs and expectations. The article emphasizes the role of User-Centered Design (UCD) in the HADRIAN project with the intent to fulfill the promise of AD. Through the iterative creation of personas, like Harold, representing the elderly, Sven, the seasoned truck driver, and Florence, the business woman, the project brings to light the requirements for AD from a user-centric focus. Such personas, backed by comprehensive research and expert insights, steer in the HADRIAN project not only the design of interfaces but influence the development of AD systems ensuring broader acceptance and inclusivity. As the landscape of urbanization and digitalization expands, coupled with the emergence of smart cities and shared mobility solutions, the integration of AD with these broader trends becomes imperative. This will require a larger ecosystem that allows to steer vehicle development processes increasingly toward the specific needs of its user groups, moving away from one-for-all vehicles. In the HADRIAN project we envisioned such an ecosystem to be able to step beyond currently prevailing vehicle development approaches and show how with a user-focused approach, the future of transportation can become a harmonious blend of safety, efficiency, and inclusivity, resonating with the real-world needs of diverse commuters.

C. Zachäus (✉)
VDI/VDE Innovation & Technik GmbH, Berlin, Germany
e-mail: carolin.zachaeus@vdivde-it.de

S. Trösterer · C. Marx · P. Moertl
Virtual Vehicle Research GmbH, Graz, Austria

P. Moertl and B. Brandstaetter (eds.), *Shaping Automated Driving to Achieve Societal Mobility Needs*, Lecture Notes in Mobility,
https://doi.org/10.1007/978-3-031-52550-6_2

Keywords User-Centred Design · User Needs · Personas · Automated Driving · Elderly · Truck Driver · Businessperson · Mobility Scenarios

1 Introduction

Recent advancements in automation have profoundly influenced a wide range of different sectors, notably the mobility domain. Highly AD promises not only enhanced operational efficiency and a potential decrease in human-induced errors but also improved safety on a broader scale. Beyond these fundamental improvements, the shift toward automation in transportation presents a suite of novel possibilities. These include innovative mobility offerings tailored for individuals previously excluded from traditional mobility paradigms, a substantial reduction in driver stress, and the opportunity to engage in more meaningful activities during transit.

However, the trajectory toward an integrative and comprehensive automated mobility ecosystem is full of challenges, encompassing both technical adaptations and broader societal considerations. One significant issue is accommodating the diverse needs of heterogeneous user groups, each of them having distinct expectations, reservations, and functional requirements.

The principles of User-Centered Design (UCD) emerge as a critical tool in this context. This methodology emphasizes the alignment of technological advancements with the intricate needs and preferences of its intended users. By emphasizing the diverse needs of different user groups, from daily commuters over professional drivers to elderly individuals or those with specific disabilities, the research and development community can facilitate a transition toward AD systems that are both technically robust and contextually relevant.

2 User-Centered Design Approach of HADRIAN

User-Centric Design (UCD) is an iterative design process that focuses on understanding users and their needs during each phase of the design process, prioritizing usability goals, user characteristics, environments and tasks (Gulliksen et al. 2003). Hereby, usability, according to DIN EN ISO 9241-11, refers to the extent to which a system, product, or service can be used by specific users in a specific context of use to achieve specific goals effectively, efficiently, and satisfactorily. It is closely related to the more colloquial term "user-friendliness" and the broader concept of User Experience (UX). The primary goal of UCD is to create products, services or processes that offer a positive and convenient user experience, which increases acceptance, usability and inclusiveness.

The UCD process generally involves four distinct phases. First, the context in which the product may be used, including the motivation of use, the environment of use and general requirements are collected. Then, more specific users' requirements are identified, followed by a design phase in which concrete solutions are developed. Finally, an evaluation phase is conducted to assess the outcomes against the users' context and requirements (Van Kujik 2010). Usually, UCD is applied to the domain of software design. However, it is becoming more and more important also to other domains even though still rarely applied to develop mobility solutions.

In the context of the HADRIAN project, the development process was adapted and structured as follows:

1. **Understanding the Users**: Extensive literature research and consultation of experts has been performed to understand the constraints, needs, and daily routines of three user groups.
2. **Persona Creation**: To foster a shared understanding of each user group and their unique characteristics, we developed representative personas. These personas, Harold (elderly driver), Sven (truck driver), and Florence (business women, working in the car), embody the key characteristics of their respective user groups.
3. **Identification of Specific Needs and Challenges**: Typical situations that each persona might encounter based on their needs and related challenges have been identified and first potential solutions have been discussed.
4. **Scenario Creation**: Mobility scenarios have been created to develop a storyline that incorporates the individual user concepts into real driving situations.
5. **Collection of Initial Requirements**: Initial requirements have been identified and further refined as a basis for the development of the HADRIAN functionalities for AD. It is important that these requirements address the specific needs of the identified personas and form effective constraints for system design, going from a generic design to a user centered one.
6. **Creation of Initial Design Solutions**: Based on the initial requirements for the selected scenarios and personas, design solutions were envisioned. These design solutions were then prototyped and evaluated with users, and based on that input, refined in subsequent iteration steps.

2.1 Understanding the Users

Understanding users in the context of connected, cooperative and automated driving is crucial for the successful development and deployment of these systems. Users come with a variety of expectations, capabilities, fears, and distinct needs that must be considered in the design process. Different user groups, such as elderly drivers, persons with disabilities, daily commuters, and commercial vehicle operators, each have their unique requirements. For instance, while some may appreciate the convenience of automation, others might be cautious about ceding control. Similarly, an elderly person may emphasize safety and user-friendliness, while a commercial driver might look for efficiency and route optimization. In the realm of connected driving, where vehicles converse with infrastructure and other vehicles, trust in the technology's reliability and security is paramount. It's essential that these systems are not just technologically advanced, but also molded to resonate with the diverse needs and concerns of their daily users. By grasping these multifaceted user perspectives, developers stand a better chance at tailoring systems that truly serve the communities they're designed for.

The understanding of the users leads to specific user groups that have general mobility needs in common, which can be met by an ADAS. Such mobility needs can be derived from European mobility visions that have been developed, for example, by the EU project Mobility4EU.[1] These mobility visions provide a larger picture of multimodal transportation over the next decades, intended to help organize and balance the various mobility developments of mass and individual transport solutions. Specifically, a review of the Mobility4EU mobility vision (Mobility4EU 2016) resulted in the identification of three specific user groups that could benefit the most from automated vehicles in the future along with their specific needs and usage situations. While these user groups are only a subset of all possible user groups in society, they have been selected, already representing a wide range of interest groups. The identified user groups are the following:

- Elderly drivers, who want to stay active after losing some perceptual, cognitive, and motor skills that are necessary for safe driving.
- Truck drivers, whose job has lost attractiveness over the last years and who experience additional driving unrelated pressures during work (time pressures, form administration, loading and unloading, etc.).
- Office workers, who want to remain productive during individual travel.

These selected user groups form a subset of European user needs for which the human-centered design and holistic solution process is developed and evaluated.

2.2 HADRIAN Personas

For the representatives of each user group, a specific persona was created that exemplifies the specific needs within the mobility context (see Fig. 1).

The term "persona" originates from Greek, meaning "mask". In the context of UCD, working with personas is aiming at gaining a deeper understanding of their needs and expectations as foundation for the development of a new product, service or process.

Alan Cooper introduced the concept of personas as a tool for enhancing user experience in software design in 1999 (Cooper 1999). Personas, as defined by Cooper, are fictional characters that represent composite archetypes, encapsulating behavioral data derived from ethnographic studies and empirical analysis of actual users. This approach allows the identification with the users, helping comprehend the users' desires and the intended use and is based on user data analysis as well as information gathered directly from real people.

[1] https://www.mobility4eu.eu/.

Fig. 1 For each of the three user groups, a specific Persona was designed to exemplify European Mobility Needs for AD: Harold, Sven, and Florence

As the use of personas gains popularity, the need for a universally defined standard becomes increasingly important. Personas are described as a model of a user class, encompassing user characteristics, goals, and needs. They are captured in narrative form, with only general guidelines currently available on how they should be represented. The primary use of personas is as a communication tool, with the aim of inspiring the design team and supporting the UCD process (Norman 1986).

Courage and Baxter (2005) define a set of persona components in a textual format, which serves as a guide to the construction of personas. These components include identity, status, goals, knowledge and experience, tasks, relationships, psychological profile and needs, attitude and motivation, expectations, and disabilities. These components can act as a guide in building personas and are refined to encapsulate user requirements.

The HADRIAN personas, shown in Fig. 1, are built on user groups and the respective user needs, requirements and expectations discussed in the paragraph above. The characteristics of the user group, including user needs, impairments, challenges as well as the derived personas are described in the following paragraphs for elderly drivers, truck drivers and office workers.

3 AD Enabling Inclusive Mobility: Example of Elderly Drivers

The purpose of driving for elderly people is to satisfy individual needs such as transportation to essential services like healthcare or groceries, independence and socialization. Musselwhite and Haddad (2018) developed a model outlining the travel needs of elderly drivers, categorized into primary, secondary, and tertiary mobility needs. Primary needs focus on practical aspects of mobility to be able to get from A to B as safely, reliably, cheaply and comfortably as possible. Secondary needs are tied to social aspects like independence and self-esteem, while tertiary needs are aesthetic, including the joy of movement and the environment. However, elderly people may have developed various impairments, particularly around the age of 60–70 years, which can impact their driving abilities. A study by Vaa (2003) assessed the relative risks of these impairments, finding cognitive impairments to be particularly dangerous, followed by cardiovascular diseases, and hearing, locomotor, and vision impairments. Besides various impairments, older people often struggle with self-assessment of their driving abilities. Horswill et al. (2011) found that elderly drivers tend to overestimate their driving capabilities, which can pose a significant safety risk.

3.1 AD for Elderly People

There is a range of strategies and concepts aimed at supporting older drivers in traffic. Davidse (2006) pinpoints the most significant factors contributing to accident rates among elderly drivers and proposes corresponding assistance ideas. The goal is to bolster the drivers' weak points without taking over tasks they are already proficient at.

As described above, potential impairments span across vision and hearing, cognitive processing and decision making and physical changes. For these impairments, a variety of ADAS have been suggested. These include automatic lane changing and merging systems to address peripheral vision issues, collision warning systems for motion perception, in-vehicle signaling systems for selective attention, and systems that provide information on complex intersections to aid speed of information processing.

The benefits of such ADAS have been proven by several studies. Dotzauer et al. (2013) demonstrated that an intersection assistant significantly improved focus on relevant aspects of the situation, leading to safer intersection crossings. Similarly, Becic et al. (2013) found that additional cognitive load resulted in a more conservative driving style in both younger and older drivers when using an intersection-crossing assistant. Li et al. (2019) conducted research on age-friendly highly automated vehicles, specifically focusing on take-over situations. Their findings indicated that providing the driver with information about the state of the vehicle and the reasons for the manual driving takeover request resulted in better takeover performance and more positive attitudes toward the vehicle. Finally, Emmerson et al. (2013) examined the impact of navigation systems on the behavior of older drivers. They found that older drivers exhibited increased confidence when using navigational systems, which also served as a form of companionship and added an element of pleasure to driving.

When it comes to the use of automated driving functions, it is not only factual safety improvements that play a role. Diepold et al. (2017) studied the acceptance of automated vehicles among older people, finding that around 75% were hesitant due to uncertainty and distrust in the technology. However, the remaining 25% were curious and willing to try it. Despite the general reluctance, older people acknowledged the potential benefits of automated vehicles, such as increased mobility and independence. Concerns were raised about security and privacy. Research by Son et al. (2015) found no significant difference in the acceptance of ADAS between younger (30–45 years) and older (60–75 years) drivers, with older drivers benefiting more from ADAS in terms of safe driving behavior. In terms of interface design, studies suggest that older people have specific needs. Kim et al. (2012) found that while younger drivers benefit from multi-modal navigation systems, older drivers can find them overwhelming. Therefore, interfaces should be personalized and simplified for older users. An interview study by Li et al. (2019) found that older drivers were open to highly AD but wanted to maintain potential control over the vehicle. They expressed the need for an adjustable and explanatory information system, and a driving style that imitates their standard driving behavior while correcting unsafe actions.

3.2 HADRIAN Persona: Harold

In the HADRIAN project the elderly user of the AD system is exemplified by Harold—a 78-year-old man living in the suburbs of Paris. He has driven his car throughout his whole life. Now, he lives alone, and his only daughter is about a 1 h drive away in the countryside. Harold, however, is hesitant to use novel technologies and does not even own a smart phone. Nevertheless, he liked the safety assistance features on his previous car.

He has recently received some driving restrictions from the local authorities due to the following impairments that had been detected in a mandatory driving test:

- Harold cannot quickly focus on relevant aspects of complicated situations under time pressure. This leads to problems at intersections.
- Harold exhausts faster than before.
- Harold has difficulties driving longer distances.
- Harold has limited peripheral vision, making it difficult for him to recognize objects coming from the side.

However, Harold can drive without problems in easy situations, such as straight roads with little traffic. Accordingly, driving on highways or long-distance driving is not safe anymore. This leads to many restrictions since he still feels vital, and wants to stay mobile.

4 AD Transforming Road Logistics: Example of Truck Drivers

Trucks are involved in about 4% of German traffic accidents. With annual external financial damages of about 393 million Euros, they are responsible for 9–16% of financial damage on German streets. Proportionally the highest financial loss takes place on motorways (Bundesanstalt für Straßenwesen 2005). The comparatively high costs of these accidents are caused by the greater value of each truck along with its potential loss of goods and the accumulated costs of a stationary truck. Truck accidents are mostly triggered by insufficient distance to the surrounding vehicles and inappropriate speed. Vehicles often cut right in front of trucks on motorways and cause truck drivers to either brake too rapidly, damage vehicles behind the truck, or directly crash into the vehicle cutting in. Incorrect lane changes are another issue in truck driving. When changing lanes on motorways, it is more likely that truck drivers overlook oncoming vehicles or losing control and crashing into the guard rail. Truck accidents also often happen at intersections. Especially vulnerable traffic participants like, e.g., bicyclists can be easily overlooked (Panwinkler 2018; Trabert et al. 2018).

The above-mentioned safety critical situations are mostly caused either by truck driver violations, where the driver intentionally breaks certain driving rules, or truck driver errors mostly due to drowsiness or distraction (Sullman et al. 2002). An important influence is thereby the work situation of truck drivers that consists of sitting in their vehicles for many hours. Truck drivers are prone to an inadequate diet, sedentary habits, and a lack of time to exercise. Additionally, dense work schedules, long work weeks, a constant feeling of being monitored, and the responsibility for the truck and its goods are straining. Consequently, many truck drivers have too little sleep and are exhausted while driving (Williams and George 2014). Many start their work-day tired. Some drivers even stated in interviews that they use long, straight roads for short naps to regenerate. Because of their demanding work routines, truck drivers are a high-risk group to develop unhealthy conditions, like cardiovascular diseases, obesity, sleep apnea, and stress (Greenfield et al. 2016). However, they do know that their unhealthy lifestyle has consequences in the long run and would like to change their lifestyle and health if the conditions would allow for that. Greenfield et al. (2016) suggest that health improvement ideas should be aligned with the unique working conditions of truck drivers.

The already mentioned issues of rule violation and error, enhanced by the feeling of boredom, manifest themselves in the form of distractions like cell phone usage while driving (Iseland et al. 2018). Too dense driving schedules may also force truck drivers to do administrative tasks and route planning while driving (Claveria et al. 2019). The safety impact of distracting secondary tasks is influenced by three dimensions: (a) The frequency of a secondary task, (b) its duration, and (c) its visual demand (Hanowski et al. 2005). Specifically, high visually demanding tasks, like the use of mobile phones when texting or browsing the internet and navigating with purely visual navigation systems, carry a very high risk. It is important to note, however, that not only visual tasks reduce visual attention on the street. A predominantly auditory stimulus does not imply that drivers will not look away from the street (Hanowski et al. 2005).

Safety risks are not limited only to cell phone usage. Secondary tasks unrelated to driving can be further categorized as work environment related necessities (e.g. getting food from the in-cabin refrigerator, eating, drinking, etc.) or administrative tasks (e.g. schedule planning, filling in logbooks, etc.).

4.1 AD for Truck Driver

The advent of AD technologies is revolutionizing the trucking industry, offering a myriad of potential benefits (Berger 2016; World Economic Forum 2021). One of the most significant improvements is in the realm of safety. AD systems can reduce the number of accidents caused by human error by maintaining safe distances from other vehicles, staying within lanes, and reacting faster than humans during emergency situations. AD systems also have the potential to alleviate driver fatigue, a significant safety risk in the trucking industry. These systems can take over the driving task, allowing drivers to rest during long trips or engage in secondary tasks such as talking to friends and family, eating and drinking or reading for leisure. This "extra" time could also be used for additional administrative tasks such as filling forms. Furthermore, automated trucks can use real-time traffic data to choose the most efficient routes and avoid congested areas, leading to improved traffic flow and optimized working hours for the truck driver. Thus, automated functionalities for truck driving contributes significantly to safety, efficiency, and cost benefits.

Besides, apparent benefits for transport safety and the truck driver itself, automated trucks could help alleviate the problem of labour shortage in the trucking industry. By reducing the industry's reliance on human drivers but also by making the profession more attractive again, automated trucks could address this issue (Berger 2016). Moreover, these systems can lead to significant fuel savings. Especially, platooning, enabled by AD technologies, can improve fuel efficiency. Automated trucks can drive at consistent speeds and make smooth transitions, significantly reducing fuel consumption. Other approaches focus on the possibilities automated trucks might offer to the driver. For example, Richardson et al. (2015) investigated the health issue of truck drivers and created a driver seat which was adaptable to the drivers' current movement and activity needs. In automated mode of the truck, it allows the driver to stand up and do exercises while safely being strapped in the seatbelt. Take-over situations, where control is handed back to the driver, are critical for safety. Truck drivers generally react faster than unprofessional drivers due to their experience. However, complex situations and distractions can delay reaction times (Lotz et al. 2019). The design of human–machine interfaces can improve reaction times by providing information about the time or distance until the next take-over. This not only reduces stress and increases control but also fosters a positive attitude towards using automated systems (Richardson et al. 2018). Furthermore, technological development towards fully automated trucks takes place in stages, in which the driver engagement changes accordingly. Each stage of automated trucks requires increasingly complex features that transfer more control from the driver to the truck (Berger 2016).

The European Union introduced the new "Vehicle General Safety Regulation" in July 2023 to mandate certain assistance systems for trucks to enhance road safety, particularly at intersections and during turns, and to mitigate accidents caused by distracted drivers (European Commission 2023). These systems, which include intelligent speed adaptation, advanced emergency braking, lane-keeping, turning assistant, and driver drowsiness and distraction warnings, are designed to monitor the vehicle's state and environment, and the driver's readiness to regain control. They will be automatically activated when the truck starts and can only be turned off through a series of actions to prevent accidental shutdowns. However, acoustic warnings can be easily silenced if they distract the driver.

Implementing automated trucks requires acceptance from operators and other stakeholders. Despite skepticism and concerns about comfort, driving pleasure, and job redundancy, research suggests that the hedonistic aspects, such as improved work-life quality, are key to acceptance (Fröhlich et al. 2018; Richardson et al. 2017). However, mistrust in technology, due to struggles with existing driving assistance systems, is a challenge (Trösterer et al. 2017). Despite this, decision-makers see potential in automated trucks to enhance safety and attract personnel in a sector facing a constant labor shortage.

4.2 *HADRIAN Persona: Sven*

Sven is a 42-year-old, long-distance truck driver living in Frankfurt. He has a wife and an 8-year-old daughter. He has been a truck driver for 20 years and is a reliable driver—that's why his boss wants to keep him at any rate, since truck drivers are hard to find nowadays. He has been almost everywhere in Europe and neighboring countries and is familiar with all the important highways. When he is on tour, he has a lot of responsibilities. He is in regular contact with his dispatcher to receive pertinent information such as his next route, any delays, etc. He must make sure that the load is secured and needs to be aware of places with prevalent criminal activity. He needs to take care of administrative items (e.g., filling in delivery forms) and must be aware of regulations regarding driving hours. Truck driving gives him a feeling of independence. Whenever he closes the door of his truck cabin, he feels like he is his own boss. Since the truck cabin is nearly like a living room for him, he takes care of it and has arranged it for his own needs. He really likes his job, even though it's not like in the past anymore because he believes that controls have increased over time. Some routes are also unpleasant to drive due to heavy traffic and numerous traffic jams. This makes his job also rather stressful. He often must fight upcoming fatigue on monotonous drives and has very long waiting times until he can load/unload, which can cause further delays. Although he likes to be independent, he also sometimes feels lonely. He misses his family who he sometimes does not see for days. So, whenever it is possible, he tries to talk to them. He is tempted to even videocall during boring, long driving stretches. Over the years, he has developed some health issues. He gained some weight due to the long hours sitting in his truck and the respective lack of exercise. He has also occasionally problems with his back, partially due to the uncomfortable bed in his truck.

5 AD Facilitating Working on Wheels: Example of Office Worker

Due to an increase in urbanization, globalization, and wealth, the individual mobility demand is constantly increasing. The global demand of passenger mobility in urbanized areas is set to double by 2050. Meanwhile, the number of individual journeys taken daily has grown massively, thereby putting increased pressure on existing mobility systems. Even larger growth is expected in the field of goods mobility, especially in dense urban areas, due to the growing importance of e-commerce and the accompanying boom in demand for last-mile delivery (Little 2018).

Smart cities are emerging, and digitalization (internet of things, big data) is constantly rising. Additionally, further societal drivers, such as environmental regulations, safety and security concerns as well as restructuring of working arrangements, are slowly changing the future of mobility.

Several mobility concepts have been developed to face the challenges ahead. Most of them rely on the interplay of electrification, automation, and sharing services to assure seamless mobility (National Association of City Transport Officials 2019; Simpson 2019). This means that sharing meets autonomy and the boundaries between private, shared, and public transport blur. Hereby, smart connected traffic management plays a major role in realizing an efficient traffic system (Mobility4EU 2016).

5.1 *AD for Office Worker*

With advancing technology, drivers are often tempted to multitask within their vehicles, diving into secondary tasks. These activities, however, aren't just about convenience; they are deeply intertwined with safety and the overall driving experience. One of the foremost considerations when engaging in secondary tasks while driving is 'interruptibility'. It is crucial to understand how engaging a task can be and how it might impact a driver's ability to promptly react if they need to suddenly assume control of the vehicle. With this knowledge and adding the duration of a task, advanced vehicles can even proactively suggest tasks that the driver can comfortably complete before any potential driving interruption. Yet, there's more to consider than just time and attention. Some tasks may pose physical demands on drivers, affecting their capacity to seamlessly shift between the activity and driving. Although secondary tasks might not dramatically delay a driver's reaction time, they can certainly influence how effectively they react. A significant challenge that comes into mind when thinking about secondary tasks in vehicles is motion sickness, or 'kinetosis'. Engaging in visually demanding tasks while in motion, such as reading or watching a movie, can lead to feelings of dizziness or nausea (Diels and Bos 2016). This dissonance between what our eyes perceive and what our bodies feel can be especially pronounced when drivers feel they have limited control over the vehicle.

Interestingly, this kind of activity plays a pivotal role in the severity of motion sickness. For instance, reading has been pinpointed as a prime culprit, while watching videos or texting follows close behind. Video gaming, surprisingly, has the least severe repercussions in this context (Sivak and Schoettle 2015).

Most concepts aim at reducing motion sickness and increasing riding comfort. AUDI "Experience Ride"[2] utilizes adapted virtual reality devices to reduce motion sickness, while BMW "I inside Future"[3] enlarges the room size of the driver's cab to increase working comfort. Within Volvo "Concept 26"[4] the mobility of the driver's seat is increased, and multiple screens are installed inside of the vehicle and within the Regus & Rinspeed "Xchange"[5] movable parts are installed to offer the driver small desks and repositories for working. Looking into design concepts to reduce motion sickness, innovative vehicle designs, like the integration of heads-up displays that keep drivers looking forward, have shown a marked reduction in motion sickness symptoms. Further design concepts focus on amplifying comfort and minimizing motion sickness by employing strategies to integrate virtual reality devices to enlarge the driver's personal space (Pretto et al. 2019). The inclusion of multiple screens and movable desks is also gaining traction, heralding a new age where office work in cars becomes not just possible, but comfortable.

5.2 *HADRIAN Persona: Florence*

Florence is a 27-year-old businesswoman living in a suburb of Paris of the future (i.e., car sharing, and intelligent traffic management are available). Florence is married and has two kids (10 months and 7 years). Her husband takes care of the kids at the moment. For a year, she led an IT start-up and has 10 co-workers. Her office is in Paris, hence she needs to commute daily. Depending on the traffic and time of the day, the drive lasts 20–35 min. Currently, she needs to take care of basically everything to ensure the success of the company. She bears a high degree of responsibility that things run smoothly as any shortfall could have severe consequences for the company. Accordingly, her workload is very high, so she tries to use every minute of her working time as efficiently as possible, even the time she spends in her car. Florence sometimes has problems reading in a moving car as it can make her feel dizzy and nauseous. Her daily tasks involve:

- Communication and information sharing with her co-workers

[2] https://www.audi-mediacenter.com/de/audi-auf-der-ces-2019-11175/audi-experience-ride-11179 assessed 04.10.2023.

[3] https://www.press.bmwgroup.com/deutschland/photo/detail/P90245530/BMW-I-Inside-Future-01-17 assessed 04.10.2023.

[4] https://www.media.volvocars.com/at/de-at/media/pressreleases/169396/neues-sitz-und-bedien konzept-volvo-concept-26-bietet-den-luxus-der-zeit assessed 04.10.2023.

[5] https://www.rinspeed.com/de/XchangE_24_concept-car.html.

- Physical and online meetings with important customers, contractors, and producers
- Administrative tasks and office work (writing, reading, calculations, internet, social media, paperwork, etc.).

6 HADRIAN Mobility Scenarios

Scenarios in the user centered design processes are stories that show what a day in the life of a user might look like, based on a persona, or character, that represents a group of users. These stories help technology designers to understand the context of the users' lives, including their emotions, challenges, and daily routines, making the user and their needs feel more real and easier to understand. This supports the understanding why users might behave in certain ways and what they expect from the product. These stories also highlight any issues or difficulties the users might face.

When creating a user scenario, it is important to consider the user's goal and the circumstances that led them to this situation, the environment, the setting of the scenario as well as the physical surroundings and any social or legal aspects. In writing these scenarios, the key is to include enough detail to make the scenario realistic and relatable, but not so much that it becomes confusing or irrelevant.

In the context of the HADRIAN project a selection of mobility scenarios was developed that exemplify typical usage situations, based on the personas and their individual mobility needs. Twelve scenarios have been extracted. A summary of these scenarios is described in Table 1. More detail about the scenarios can be found in the project deliverable.[6]

7 Initial Iteration of User-Centered Design Innovations

Based on the individual constraints and needs of the identified personas in their specific scenarios, the consortium developed a set of initial design solutions that were investigated further and subsequently refined. As identified in the introduction, the term design does here not only refer to the design of the human-vehicle interface but to the functionality of a complete vehicle system that includes the functions and also the interfaces. These solutions are depicted in Fig. 2 and are described further in the remainder of this book.

1. ***Environment awareness assistant***: The environment awareness assistant provides critical driving related information to the driver in manual driving mode and is intended to help compensate some driving skill deteriorations for continued

[6] https://hadrianproject.eu/wp-content/uploads/2020/10/HADRIAN_D_1.1.pdf.

Table 1 Mobility scenarios

No	ID	Persona and trip purpose	Environment
1	H1	Harold, an elderly driver, on a countryside trip to visit his daughter	Countryside highway
2	H2	Harold driving toward his vacation destination	Motorway
3	H3	Harold driving to his doctor	Urban
4	S1	Sven, a truck driver, driving under stress	Motorway
5	S2	Sven driving in monotonous traffic	Motorway
6	S3	Sven driving on multi-day long distance trip	Motorway
7	F1	Florence, driving to work, doing light office work	Urban and suburban
8	F2	Florence, driving to work, doing intense office work	Urban and suburban
9	F3	Florence, driving to work, disengaging extensively from driving	Urban and suburban
10	F4	Florence leasing an automated vehicle	Urban and suburban

Fig. 2 The HADRIAN innovations for different modes of automated driving

safe driving. Specifically, the assistant should compensate some of the deterioration of perceptual or cognitive driving skills of elderly drivers and thereby help them to keep feeling useful, active, and engaged in their daily lives. Such information is provided via auditory, visual, or haptic means, for example, by highlighting a crossing pedestrian on the windshield, giving haptic cues on the steering wheel, or supporting situation awareness via auditory channel.

2. ***Warning prior to disengagement of ADL 2***[7]: The driver must perform two different types of monitoring tasks during ADL 2. First, the driver must monitor the driving automation and respond to any disengagements with the appropriate

[7] Based on the SAE classification of automated driving levels (SAE International 2021), partial automated driving (level 2) requires the driver to supervise the automated driving system at all times. See Chap. 1 of this book for potential human factors problems.

control actions. Second, the driver must monitor the environment for objects or events that may be safety relevant and require a manoeuvre. Current implementations of ADL 2 in vehicles provide no advanced warnings for such situations, the driver is only warned at the time of the ADL 2 disengagement. This can deteriorate safety as was discussed in Chap. 1. Therefore, this HADRIAN innovation provides drivers a 5 s warning prior to the disengagement of ADL 2 to help them taking active control again.

3. ***Guaranteed transition time from ADL 3 back to manual driving***[8]: As was discussed in Chap. 1, taking back active control after periods of disengagement can be challenging for drivers. Therefore, this innovation guarantees drivers at least 15 s of time to respond to a request to intervene. The duration of 15 s was selected as the 10 s that are required for automated lane keeping systems (UN Regulation No. 167, Addendum 166, 2023) were found to be insufficient in many situations (Shi and Bengler 2022). However, more important than the amount of time is that the duration is guaranteed to the drivers through the road information infrastructure which sends timely safety critical road information to the vehicle. This guaranteed minimum duration should make the take-over predictable to drivers and help them learn to perform them safely through training and repetition.
4. ***Predictable ADL 3 duration***: One primary expected benefit of driving at ADL 3 is that it allows drivers to engage in other entertaining or productive non-driving related activities (NDRA), while being driven safely by the automation. However, to be able to practically use such periods of leisure or work, drivers would have to know in advance whether and for how long driving at ADL 3 would be available. Some NDRA may be useful for drivers only when they can be performed for some extended period of time such as writing a report. Therefore, telling drivers in advance about the expected duration of their ADL 3 portion of the trip should increase the usefulness of ADL 3. It may also have a positive safety impact as drivers have information to complete their NDRA prior to the transition back to manual driving.
5. ***Fluid interactions***: As was outlined in Chap. 1, fluid interactions consist of the vehicle to adapt to the perceived state and behavior of the driver and occupant and offer information only when and how it is needed. This should, if appropriately designed, help reduce the amount of effort and knowledge that drivers require to appropriately adjust to the specific functions and operations of their automated vehicle. To reduce the amount of adaptation that drivers must perform, fluid interactions shift adaptations to the vehicle; the vehicle monitors the state and behavior of the driver and provides aiding information or intervenes in cases when needed. For example, the vehicle checks whether the driver does not sufficiently

---

[8] Based on the SAE classification of automated driving levels (SAE International 2021), conditional automated driving (level 3) requires the driver to remain fallback ready while the automated driving system takes over lateral and longitudinal maneuvers and requests the driver to intervene when needed. See Chap. 1 of this book for potential human factors problems.

monitor the vehicle and situation during ADL 2 or performs incorrect take-over maneuvers in ADL 3.

6. ***Driver driving assistence system (DAS) tutoring application***: It is current practice that drivers learn how to use the DAS by trial-and-error, over time learning to understand how to safely use the DAS. Such trial-and-error learning can result in unsafe driving states when the driver is surprised about a previously not experienced limitation of the driving assistance (e.g., unexpected disengagement of the lane-keeping assistant). Furthermore, every driver would have to go through such a trial-and-error period and may still only know the limitations within their commonly used driving context. The driver DAS tutoring application helps the driver acquire the skills and competences necessary to safely use the DAS before, during, and after the drive.
7. ***The guardian angel***: This innovation specifically addresses driving performance degradations of elderly drivers: the guardian angel monitors manual driving and intervenes through longitudinal or lateral control maneuvers, if needed, to bring the vehicle again into a safe state. This innovation is described in further detail in Chap. 4.

These initial design solutions were subsequently implemented and investigated in the HADRIAN project and results of these investigations and refinements are described in the subsequent sections in this book. It is important to note that the investigated design solutions as they are described above are not on solely on the interaction level but formulated on a higher level of abstraction. For example, fluid interfaces as adaptive automation, can be understood as a concept rather than a specific interface solution. Also, "AD predictability" is not an interface feature but a basic underlying function of the DAS. These initial design solutions can be implemented with different interfaces, but before reaching these points it needs to be shown whether the functions per se provide benefits in terms of safety, comfort, and acceptance. For this purpose, the design prototypes that were implemented in the HADRIAN project were mainly intended for research to confirm or disconfirm the underlying principles rather than to confirm the interface implementations. Different vehicle vendors would certainly have their own ideas and guidelines to create such interface designs based on their often mode-specific cabin design guidelines.

8 Conclusion

The paradigm shift toward AD technologies in the mobility sector presents an unprecedented opportunity to revolutionize transport efficiency, safety, and user experience. As we delve into the complexities of this transformation, it becomes clear that the challenges are as varied as the users themselves. From the specific needs of elderly drivers to the unique challenges faced by truck operators, a one-size-fits-all approach is both unrealistic and ineffective.

The HADRIAN project's utilization of UCD shows how a user-centric approach can be adopted by vehicle developers to meet the new requirements for automated vehicles that are used very differently than traditional manually driving vehicles. For example, automated vehicles can be effectively used within the context of mobility as service where vehicles are designed to meet the needs of specific user groups, thereby allowing for optimized shared use by many people.

By crafting personas such as Harold, Sven and Florence, the project underscores the need to understand, respect, and design for the diverse requirements and apprehensions of different user groups. As urbanization continues apace and intertwines with technological advancements like smart cities and shared mobility, the need for seamless integration becomes even more pressing. Embracing UCD in the development of AD systems ensures that as we move toward this automated future, we are guided by a compass that prioritizes inclusivity, resonance with real-world needs, and the broad acceptance of these emerging technologies.

Doing so, it is important to remember that the kind of ecosystem that allows to bring together the mobility needs of groups of people with the forces to build such systems may not yet exist. Vehicle manufacturers traditionally still build general purpose vehicles that are sold to individual buyers. However, this is changing rapidly as automated driving enables quite different use cases. This provides an emerging marketplace for different types of vehicles from the one being built to date: those that are tailored to specific user and mobility needs.

The HADRIAN project envisioned such a larger ecosystem to enable vehicle development outside traditional frameworks and apply user-centered development not only to the design of the interface functions but to the design of functions addressing the entire vehicle system. Doing so will be a requirement in the future and is already reflected by systems engineering efforts to support human-systems integration as next step of UCD.[9]

Acknowledgements The human-centered design approach that was described in this chapter was the joint effort of many HADRIAN partners. Therefore, we want to sincerely thank the following people for their contributions: Christoph Pratt (CEA); Jaka Sodnik, Kristina Stojmenova, Klemens Novak (Nervtech); Alexander Mirnig, Hanna Braun (PLUS); Evelyn Gianfranchi, and Leandro DiStasi (UGR). HADRIAN has received funding from the European Union's Horizon 2020 research and innovation programme under grant agreement No 875597. This document reflects only the author's view, the Climate Innovation and Networks Executive Agency (CINEA) is not responsible for any use that may be made of the information it contains.

[9] https://www.incose.org/incose-member-resources/working-groups/analytic/human-systems-integration.

References

Becic E, Manser M, Drucker C, Donath M (2013) Aging and the impact of distraction on an intersection crossing assist system. Accid Anal Prev 50:968–974. https://doi.org/10.1016/j.aap.2012.07.025

Berger R (2016) Automated trucks—the next big disrupter in the automotive industry? Roland Berger study. https://www.rolandberger.com/publications/publication_pdf/roland_berger_automated_trucks_20160517.pdf. Last assessed 01 Sept 2023

Bundesanstalt für Straßenwesen (Hrsg.) (2005) Kolloquium „Mobilitäts-/Ver-kehrserziehung in der Sekundarstufe". Wirtschaftsverlag NW

Claveria JB, Hernandez S, Anderson JC, Jessup EL (2019) Understanding truck driver behavior with respect to cell phone use and vehicle operation. Transport Res F: Traffic Psychol Behav 65:389–401. https://doi.org/10.1016/j.trf.2019.07.010

Cooper A (1999) The inmates are running the asylum. In: Arend U, Eberleh E, Pitschke K (eds) Software-ergonomie'99. Berichte des German Chapter of the ACM, vol 53. Vieweg+Teubner Verlag, Wiesbaden. https://doi.org/10.1007/978-3-322-99786-9_1

Courage C, Baxter K (2005) Understanding your users: a practical guide to user requirements methods, tools, and techniques. Elsevier, San Francisco, CA

Davidse RJ (2006) Older drivers and ADAS. IATSS Res 30(1):6–20. https://doi.org/10.1016/S0386-1112(14)60151-5

Diels C, Bos JE (2016) Self-driving carsickness. Appl Ergon 53:374–382. https://doi.org/10.1016/j.apergo.2015.09.009

Diepold K, Götzl K, Riener A, Frison A-K (2017) Automated driving: acceptance and chances for elderly people. In: Proceedings of the 9th international conference on automotive user interfaces and interactive vehicular applications adjunct—AutomotiveUI'17, pp 163–167. https://doi.org/10.1145/3131726.3131738

Dotzauer M, Caljouw SR, de Waard D, Brouwer WH (2013) Intersection assistance: a safe solution for older drivers? Accid Anal Prev 59:522–528. https://doi.org/10.1016/j.aap.2013.07.024

Emmerson C, Guo W, Blythe P, Namdeo A, Edwards S (2013) Fork in the road: in-vehicle navigation systems and older drivers. Transport Res F: Traffic Psychol Behav 21:173–180. https://doi.org/10.1016/j.trf.2013.09.013

European Commission (2023) European road safety observatory: road safety thematic report—professional drivers of trucks and buses

Fröhlich P, Sackl A, Trösterer S, Meschtscherjakov A, Diamond L, Tscheligi M (2018) Acceptance factors for future workplaces in highly automated trucks. In: Proceedings of the 10th international conference on automotive user interfaces and interactive vehicular applications—AutomotiveUI'18, pp 129–136. https://doi.org/10.1145/3239060.3240446

Greenfield R, Busink E, Wong CP, Riboli-Sasco E, Greenfield G, Majeed A, Car J, Wark PA (2016) Truck drivers' perceptions on wearable devices and health promotion: a qualitative study. BMC Public Health 16(1):677. https://doi.org/10.1186/s12889-016-3323-3

Gulliksen J, Göransson B, Boivie I, Blomkvist S, Persson J, Cajander A (2003) Key principles for user-centred systems design. Behav Inform Technol 22(6):397–409. https://doi.org/10.1080/01449290310001624329

Hanowski RJ, Perez MA, Dingus TA (2005) Driver distraction in long-haul truck drivers. Transport Res F: Traffic Psychol Behav 8(6):441–458. https://doi.org/10.1016/j.trf.2005.08.001

Horswill MS, Anstey KJ, Hatherly C, Wood JM, Pachana NA (2011) Older drivers' insight into their hazard perception ability. Accid Anal Prev 43(6):2121–2127. https://doi.org/10.1016/j.aap.2011.05.035

Iseland T, Johansson E, Skoog S, Dåderman AM (2018) An exploratory study of long-haul truck drivers' secondary tasks and reasons for performing them. Accid Anal Prev 117:154–163. https://doi.org/10.1016/j.aap.2018.04.010

Kim S, Hong J-H, Li KA, Forlizzi J, Dey AK (2012) Route guidance modality for elder driver navigation. In: Kay J, Lukowicz P, Tokuda H, Olivier P, Krüger A (eds) Pervasive computing,

edn 7319. Springer Berlin Heidelberg, pp 179–196. https://doi.org/10.1007/978-3-642-31205-2_12
Van Kujik J (2010) Recommendations for usability in practice (or how I would do it). https://studiolab.ide.tudelft.nl/studiolab/vankuijk/files/2011/11/Recommendations_Usability_Practice_van_Kuijk.pdf. Last assessed 01 Sept 2023
Li S, Blythe P, Guo W, Namdeo A (2019) Investigation of older drivers' requirements of the human-machine interaction in highly automated vehicles. Transp Res F: Traffic Psychol Behav 62:546–563. https://doi.org/10.1016/j.trf.2019.02.009
Little DA (2018) Future of mobility. https://www.adlittle.com/sites/default/files/viewpoints/adl_uitp_future_of_mobility_3.0_1.pdf. Last assessed 01 Sept 2023
Lotz A, Russwinkel N, Wohlfarth E (2019) Response times and gaze behavior of truck drivers in time critical conditional automated driving take-overs. Transp Res F: Traffic Psychol Behav 64:532–551. https://doi.org/10.1016/j.trf.2019.06.008
Mobility4EU (2016) Societal needs and requirements for future transportation and mobility as well as opportunities and challenges of current solutions. https://www.mobility4eu.eu/?wpdmdl=1245
Musselwhite C, Haddad H (2018) Older people's travel and mobility needs: a reflection of a hierarchical model 10 years on. Qual Ageing Older Adults 19(2):87–105. https://doi.org/10.1108/QAOA-12-2017-0054
National Association of City Transport Officials (2019) Blueprint for automonous urbanism, 2nd edn
Norman DA (1986) Cognitive engineering. In: Norman DA, Draper SW (eds) User centred system design: new perspectives on human computer interaction. Hillsdale N.J., Elbraum, pp 31–61
Panwinkler T (2018) Unfallgeschehen schwerer Güterkraftfahrzeuge. Berichte der Bundesanstalt fuer Strassenwesen. Unterreihe Mensch und Sicherheit, p 277
Pretto P, Mörtl P, Neuhuber N (2019) Fluid interface concept for automated driving
Richardson NT, Flohr L, Michel B (2018) Takeover requests in highly automated truck driving: how do the amount and type of additional information influence the driver-automation interaction? Multimodal Technol Interact 2(4):68. https://doi.org/10.3390/mti2040068
Richardson NT, Sinning M, Fries M, Stockert S, Lienkamp M (2015) Highly automated truck driving: how can drivers safely perform sport exercises on the go? In: Adjunct proceedings of the 7th international conference on automotive user interfaces and interactive vehicular applications—AutomotiveUI'15, pp 84–87. https://doi.org/10.1145/2809730.2809733
Richardson N, Doubek F, Kuhn K, Stumpf A (2017) Assessing truck drivers' and fleet managers' opinions towards highly automated driving. In: Stanton NA, Landry S, Di Bucchianico G, Vallicelli A (eds) Advances in human aspects of transportation, edn 484. Springer International Publishing, pp 473–484. https://doi.org/10.1007/978-3-319-41682-3_40
Shi E, Bengler K (2022) Non-driving related tasks' effects on takeover and manual driving behavior in a real driving setting: a differentiation approach based on task switching and modality shifting. Accid Anal Prev 178:106844. https://doi.org/10.1016/j.aap.2022.106844
Simpson C (2019) Mobility 2030: transforming the mobility landscape. https://assets.kpmg.com/content/dam/kpmg/xx/pdf/2019/02/mobility-2030-transforming-the-mobility-landscape.pdf. Last assessed 01 Sept 2023
Sivak M, Schoettle B (2015) Motion sickness in self-driving vehicles. 15
Son J, Park M, Park BB (2015) The effect of age, gender and roadway environment on the acceptance and effectiveness of Advanced Driver Assistance Systems. Transport Res F Traffic Psychol Behav 31:12–24. https://doi.org/10.1016/j.trf.2015.03.009
Sullman MJM, Meadows ML, Pajo KB (2002) Aberrant driving be-haviours amongst New Zealand truck drivers. Transp Res F: Traffic Psychol Behav 5(3):217–232. https://doi.org/10.1016/S1369-8478(02)00019-0
Trabert T, Shevchenko I, Müller G (2018) In-depth analyse schwerer Unfälle mit schweren Lkw. Forschungsbericht/Unfallforschung der Versicherer (GDV) 54

Trösterer S, Meneweger T, Meschtscherjakov A, Tscheligi M (2017) Transport companies, truck drivers, and the notion of semi-autonomous trucks: a contextual examination. In: Proceedings of the 9th international conference on automotive user interfaces and interactive vehicular applications adjunct—AutomotiveUI'17, pp 201–205. https://doi.org/10.1145/3131726.3131748

United Nations (2021) Regulation 157

Vaa T (2003) Impairments, diseases, age and their relative risks of accident involvement: results from meta-analysis. Transportøkonomisk institutt

Williams AJ, George BP (2014) Truck drivers—the under-respected link in the supply chain: a quasi-ethnographic perspective using qualitative appreciative inquiry. Oper Supply Chain Manage Int J 85. https://doi.org/10.31387/oscm0150093

World Economic Forum (2021) Autonomous trucks: an opportunity to make road freight safer, cleaner and more efficient—white paper. https://www3.weforum.org/docs/WEF_Autonomous_Vehicle_Movement_Goods_2021.pdf. Last assessed 01 Sept 2023

An Integrated Display of Fluid Human Systems Interactions

Sandra Trösterer, Cyril Marx, Nikolai Ebinger, Alexander Mirnig, Grega Jakus, Jaka Sodnik, Joseba Sarabia Lezamiz, Marios Sekadakis, and Peter Moertl

Abstract Supporting drivers in different levels of automation was one of the key goals in the HADRIAN project. Following the approach of a "fluid interface", i.e., an interface that considers the state of the driver, vehicle, and environment, and which uses different modalities to support the driver, a human–machine interface (HMI) was developed and compared to a baseline HMI in a simulator study (n = 39). The integrated fluid HMI aimed at supporting the driver in automated driving in SAE level 2 and 3 by providing mode relevant information, supporting the driver during take-over requests by the system, and supporting engagement in non-driving related tasks when allowed. The fluid HMI consisted of several components (head-up display, LEDs, haptic icons, sound, tablet) and featured driver monitoring and adaptive tutoring. Study results did not show significant differences between the HMIs regarding subjective measures such as user experience, usability, acceptance, or safety feeling. The various factors contributing to this conclusion are thoroughly discussed. However, objective measures in terms of eye movements and a safety analysis including driving data showed a significant benefit of the integrated fluid HMI over the baseline HMI. Participants had better mode awareness and a higher safety score with the integrated fluid HMI. Furthermore, valuable insights on how to further improve the HMI could be gained during the study.

S. Trösterer (✉) · C. Marx · N. Ebinger · P. Moertl
Virtual Vehicle Research GmbH, Graz, Austria
e-mail: sandra.troesterer@v2c2.at

A. Mirnig
Center for Human-Computer Interaction, Salzburg, Austria

G. Jakus · J. Sodnik
University of Ljubljana, Ljubljana, Slovenia

J. S. Lezamiz
Tecnalia Corporación Tecnológica, Derio, Spain

M. Sekadakis
National Technical University of Athens, Athens, Greece

P. Moertl and B. Brandstaetter (eds.), *Shaping Automated Driving to Achieve Societal Mobility Needs*, Lecture Notes in Mobility,
https://doi.org/10.1007/978-3-031-52550-6_3

Keywords Automated driving · Human machine interface · Fluidity · Driver monitoring · Tutoring · Driver distraction

1 Introduction

According to the SAE levels of driving automation (SAE International 2021), the vehicle supports the driver in steering and braking/accelerating in automated driving level 2 (ADL2), e.g., with lane keeping assistance and adaptive cruise control activated. However, the driver is still required to monitor and intervene if necessary. In automated driving level 3 (ADL3), the vehicle fully takes over the driving task for some periods of time. The driver can engage in other activities during those periods but must take over if the system is not able to perform the driving task anymore. And therein lies the challenge, since the driver has different responsibilities and options for what to do depending on the automation driving level. A misunderstanding or misconception by the driver can have a negative impact on safety. Hence, how to make drivers aware of what they can and cannot do in the respective driving level, is a crucial question.

In the HADRIAN project, we aimed at answering this question by developing an HMI with several key functionalities and pursuing the approach of fluidity. The integrated fluid HMI primarily targets the needs of the persona "Florence" as described in the chapter "User-Centered Design of Automated Driving to Meet European Mobility Needs", i.e., to support office work while driving in ADL2 or ADL3. In this chapter, we present the integrated fluid HMI, which has been developed within the HADRIAN project, and the results of an experimental study that compared this HMI to a baseline HMI.

Pretto et al. (2020) state that a fluid interface "*consists of visual, auditory, and haptic displays, allowing information to 'flow' across different sensory modalities and around the driver, adapted to his/her current activity and focus of attention*". Furthermore, such interfaces "*continuously adapt to the human operator depending on the changes in the configuration between driver, vehicle and environment*". The expectation is that a fluid interface should outperform a conventional one, because it better adapts to the driver's needs by providing relevant information when it's needed and in a timely manner.

In conceptualizing the interface, we took those points into account by *integrating* different interfaces (visual, auditory, haptic), which have been developed and explored by individual HADRIAN project partners. *Fluidity* was addressed in different regards by adapted timings, tailored tutoring based on driver behaviour, and adapted distraction warnings. A detailed overview of the HMI and its functions is provided in a following section.

The main goal of the experimental study was to evaluate whether the integrated HADRIAN fluid HMI is safer, more acceptable, more useful, and more comfortable for the driver compared to a baseline HMI. Particularly, we were interested in finding out whether the features of automated driving predictability, monitoring, and tutoring provide a benefit for the driver.

2 Related Work

As outlined, the development of the HMI was built on previous work in HADRIAN. In the following, we provide a brief overview of research associated with predictability, tutoring, ambient light, haptic icons, and the respective outcomes of previously conducted exploratory studies.

2.1 Predictability

Marx et al. (2022) investigated whether and how it benefits the driver when they are provided with information about the duration of an automated drive and the available time for a take-over. In a simulator study, they compared different levels of this information, which was presented visually: takeover time only, automated driving duration only, and a combination of the takeover and automated driving time. They found that the combined information was perceived as highly useful by the participants and led to a safer gazing behaviour of participants over time, i.e., participants monitored the road environment more during takeovers compared to the other conditions. The HADRIAN fluid HMI took these results into account by providing the combined time information.

2.2 Tutoring

While driving an automated vehicle for the first time, drivers learn using automated functions by trial and error, and thus have insufficient skills and understanding. Car owners use the owner's manual as their primary source of information to learn about the driving assistant functions (Boelhouwer et al. 2020a). However, reading written information about partial driving automation does not lead to safer use (Boelhouwer et al. 2019). Consequently, drivers try out different interaction strategies while driving (Strand et al. 2011) or have an insufficient understanding (DeGuzman and Donmez 2021).

Providing drivers with tutoring applications can prepare them better for automated driving than owner's manuals. Existing tutoring approaches include pre-drive information (Ebnali et al. 2021; Forster et al. 2019; Neuhuber et al. 2022; Payre et al. 2017) and information while driving (Boelhouwer et al. 2020b; Neuhuber et al. 2022; Rukonić et al. 2022). Research shows that tutoring improved the use of automation, take-overs, and trust calibration (cf. Boelhouwer et al. 2020b; Ebnali et al. 2021; Payre et al. 2017; Neuhuber et al. 2022).

Ebinger et al. (2023) investigated a tutoring approach consisting of a pre-drive video, auditory explanations while driving, and adaptive feedback during takeovers for ADL3 driving in a simulator study. Compared to a baseline with written instructions only, the tutoring improved drivers' mental model and attention allocation during takeovers. This tutoring approach has been used as a basis for the HADRIAN fluid HMI and was further adjusted and adapted to ADL2 and ADL3 driving.

2.3 Ambient Light

Information on the current driving or automation mode is usually displayed in the common in-vehicle interface areas (central cluster, HUD, etc.) alongside the other standard indicators regarding vehicle functions and status (Wang et al. 2020). In order to prevent mode confusion, i.e., an incorrect belief or perception of the vehicle's (current) mode of automation, additional indicators and/or different modalities (Revell et al. 2021; Koo et al. 2016; Wintersberger et al. 2019) are being explored to provide clarity beyond the standard visual indicators (Dönmez Özkan et al. 2021). Ambient lights have been used to provide mode information in the peripheral field of vision, positioned at the bottom of the windshield, along the a-pillars, or on the steering wheel (Löcken et al. 2016; Hecht et al. 2022) with the intention of acting as a supplementary display to the primary visual in the vehicle. For the HADRIAN fluid HMI, we have adapted this approach to provide assistive information on driving mode(s) via ambient light cues.

Mirnig et al. (2023) conducted a study in which the participants rated a range of colours across three dimensions: perceived degree of vehicle automation, driver engagement, and urgency. The colour set consisted of white, yellow, orange, red, green, blue, cyan, and magenta and was based on colours described in the ISO 3864-4 (International Organization for Standardization 2011) and ANSI Z535 (National Electrical Manufacturers Association 2022) standard. Cyan was added due to its use in contemporary vehicles and demonstrators to denote automation systems or automated functionalities. Both cyan and blue received high ratings for vehicle automation (with cyan being rated highest) and low ratings for both driver engagement and urgency. For the HADRIAN fluid HMI, we thus chose cyan to denote ADL3 and blue for ADL2. Yellow, orange, and red were ordered respectively with respect to increased urgency and were used as warning and escalation colours, independent of the driving mode.

2.4 Haptic Icons

Sarabia et al. (2022) investigated haptic feedback to foster the understanding of actions of automation while driving. They implemented different haptic icons at the steering wheel to convey different types of information to the driver. A set of informative icons (non-urgent notifications), transition icons (for ADL2 activation/deactivation), and take-over request icons were investigated in a simulator study in terms of their ability to support the driver. The icons chosen for the HADRIAN fluid HMI were based on the study results and represent the most feasible icons for transitions and take-over requests.

3 The HADRIAN Integrated Fluid HMI

The main aim of an integrated fluid HMI was (1) to support the driver while driving manually (in ADL2) and ADL3 by providing mode relevant information, feedback, and warnings, thus, increasing mode awareness, (2) to support drivers in case of a take-over request by the system, and (3) to allow drivers to safely engage in non-driving related tasks (NDRA) when it is allowed.

To achieve these aims, we implemented and integrated selected innovations developed within the HADRIAN project, thereby considering the findings from the previously mentioned studies focusing on these innovations. Figure 1 provides an overview of the key innovations implemented in the HADRIAN integrated fluid HMI.

The first innovation shall provide a better *automated driving predictability, availability, and continuity*. The key feature is a granted take-over time of 5 s for ADL2 and 15 s for ADL3. For ADL3 the duration can be predicted through road infrastructure integration, i.e., by using infrastructural information to determine for how long ADL3 driving would be possible for certain road segments. With *driver monitoring* (i.e., by tracking eye movements to determine visual distraction, and tracking hands-off-wheel with a camera) it shall be ensured that unsafe driver states are detected. *Tutoring* before the drive (tutoring video) and during the drive (through verbal guidance and feedback and the possibility to repeat tutoring video chapters in ADL3) aims at teaching the driver how to better use the automated driving functions

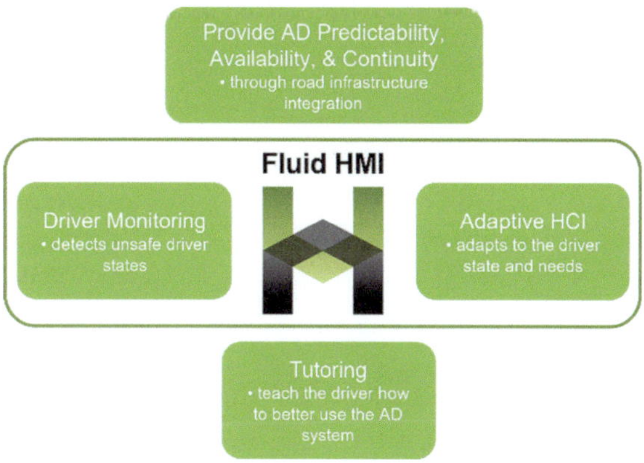

Fig. 1 Key features of the HADRIAN integrated fluid HMI

and driver's responsibilities in different driving modes. This all adds to an *adaptive human–computer interface*, which adapts to the driver states and needs by providing tailored information, feedback, and alerts.

Hence, the HADRIAN integrated fluid HMI encompassed the following key functionalities:

- 5 s time for take-overs in ADL2, and 15 s time for take-overs in ADL3: the countdown information is displayed to the driver
- Ensured time interval in which ADL3 driving is available: the duration is displayed to the driver
- Tutoring video before the drive, outlining the driving functions (i.e., driving assistance and driving automation), correct system use, and driver responsibilities in the different driving modes
- Verbal guidance and feedback for system use and adaptive tutoring during the drive based on gazing behavior and take-over time during a take-over
- Warnings in case of detected visual distraction or hands-off-wheel during manual or ADL2 driving with times adapted to the driving mode (see Table 1)

Table 1 Captured data and real-time use of data for the integrated fluid HMI

Data Source	Captured data (real-time use of data in HADRIAN HMI)
Eye-Tracking System SmartEye 60Hz with 2 cameras	Eye-movements *Trigger for distraction warning* • *Gaze off road > 2 s (manual)* • *Gaze off road > 3 s (ADL2)* *Trigger for adaptive tutoring for take-overs*
Simulator environment created with Carmaker	Driving data *Speed, Speed limit, calculation of ADL3 duration* *Trigger for showing upcoming maneuvers of the car in ADL3* *Trigger for ADL2, ADL3, and TORs*
Camera	Video data for hands-off-wheel detection *Trigger for hands-off-wheel warning* • *Both hands off wheel > 1 s (manual)* • *Both hands off wheel > 3 s (ADL2)*
Pre-questionnaire LimeSurvey	Demographic data, affinity for technology interaction (ATI) Scale (Franke et al. 2019), pre-experience with ADAS
Questionnaire LimeSurvey	UEQ (Laugwitz et al. 2008), selected CTAM subscales (Osswald et al. 2012), NASA-TLX (Hart and Staveland 1988), SART (Selcon and Taylor 1990), Trust scale (Neuhuber et al. submitted), mental model questions
Final interview	Impressions of about the system, helpful/less helpful system components, suggestions for improvements, etc

- Take-over support by providing countdown-information, current speed limit, and information about upcoming obstacles, as well as haptic cues at the steering wheel
- Information about vehicle behavior in ADL3 (current speed, detected speed limit, upcoming maneuvers)
- Support of mode awareness, mode changes, and warnings with ambient lighting

In the design of the integrated fluid HMI, we addressed these functionalities by implementing and integrating different interfaces (each provided and explored by different project partners), which serve a specific purpose within the HMI. Figure 2 provides an overview of the components and the setup in the driving simulator used for the experimental study.

HMI tablet and sound: the tablet (mounted in the upper area of the center stack) was primarily used for *providing tutoring information* (i.e., video before the drive, verbal feedback during the drive), *countdown information* in case of a take-over request, and *warning messages* (in case of detected hands-off-wheel or visual distraction). In ADL3, the remaining time in this mode was indicated on the tablet, as well as the current speed limit, speed, and upcoming driving maneuvers of the vehicle, i.e., the *vehicle behavior in ADL3*. Different sounds would indicate the (1) availability of a driving function, their (2) activation and (3) deactivation, and a (4) take-over request.

Head-up display (HUD): the main purpose of the HUD was to show *information relevant for the driving task* (i.e., current speed and detected speed limit, symbol for ADL2 or ADL3 if activated, see Fig. 3b, e), to *support the take-over* (by showing

Fig. 2 Components of the HADRIAN integrated fluid HMI and setup in the driving simulator

take-over request messages, countdown-information, obstacles/closed lanes ahead during a take-over, see Fig. 3f), and to *provide warning messages* (in case of detected hands-off-wheel or visual distraction).

Ambient light: The main purpose of the ambient light was to support mode awareness. LEDs were used at two different positions: (1) An LED stripe was mounted at the bottom of the screen showing the driving simulation. The main purpose of these LEDs was to indicate the *availability of a certain driving mode* (i.e., by flashing in blue for ADL2 or in cyan for ADL3), to indicate the *activated driving mode* (i.e., constant blue illumination for ADL2 and cyan illumination for ADL3, see Fig. 3b, e), and to warn the driver in case of detected hands-off-wheel or distraction (i.e., center part of LED stripe flashing yellow to draw attention, see Fig. 3c, d). (2) In ADL3, LEDs mounted behind the NDRA tablet would indicate that it is allowed to perform the NDRA by magenta illumination (see Fig. 3e).

Haptic cues on the steering wheel: The main purpose of the haptic cues (i.e., different vibration patterns) was to support the take-over and hand-over by indicating the availability of assisted or automated driving, the activation/deactivation of a driving mode, and by providing countdown information through increasing amplitude of the vibrations during a TOR.

As outlined, **eye-tracking** data, **hands-off-wheel** detection, as well as **road infrastructure data** like, e.g., closed lanes or the possibility to drive in ADL3 for certain road segments (in the study provided by the simulation), informed the HMI and therefore can be seen as further components of the fluid HMI.

4 Baseline HMI

In the experimental study, we compared the performance and feedback of human drivers when using either the HADRIAN fluid HMI or a baseline HMI (see Fig. 4).

For a fair comparison, we assumed the same take-over times for the HADRIAN HMI and the baseline HMI for ADL2 and ADL3. These were 5 s for ADL 2 and 15 s for ADL3. However, in the baseline HMI, these times were not indicated (outlined in the written manual, though). Also, since it is common in modern cars with driving assistance, a hands-off-wheel warning would be provided when driving in ADL2. Similar to the fluid HMI, a warning would be issued if the driver put his hands off the wheel for more than three seconds (cf. Table 1). Other than that, no adaption to the driver state was provided and no tutoring was available.

The baseline HMI consisted of the following components:

HMI tablet and sound: The tablet indicated the availability and activation of ADL2 and ADL3 by showing the respective symbols, and take-over request messages. Similar to the fluid HMI, different sounds would indicate the availability of a driving function, their activation/deactivation, and a take-over request.

Head-up display (HUD): The main purpose of the HUD was to show information relevant for driving (i.e., speed and speed limit), and to indicate the availability and

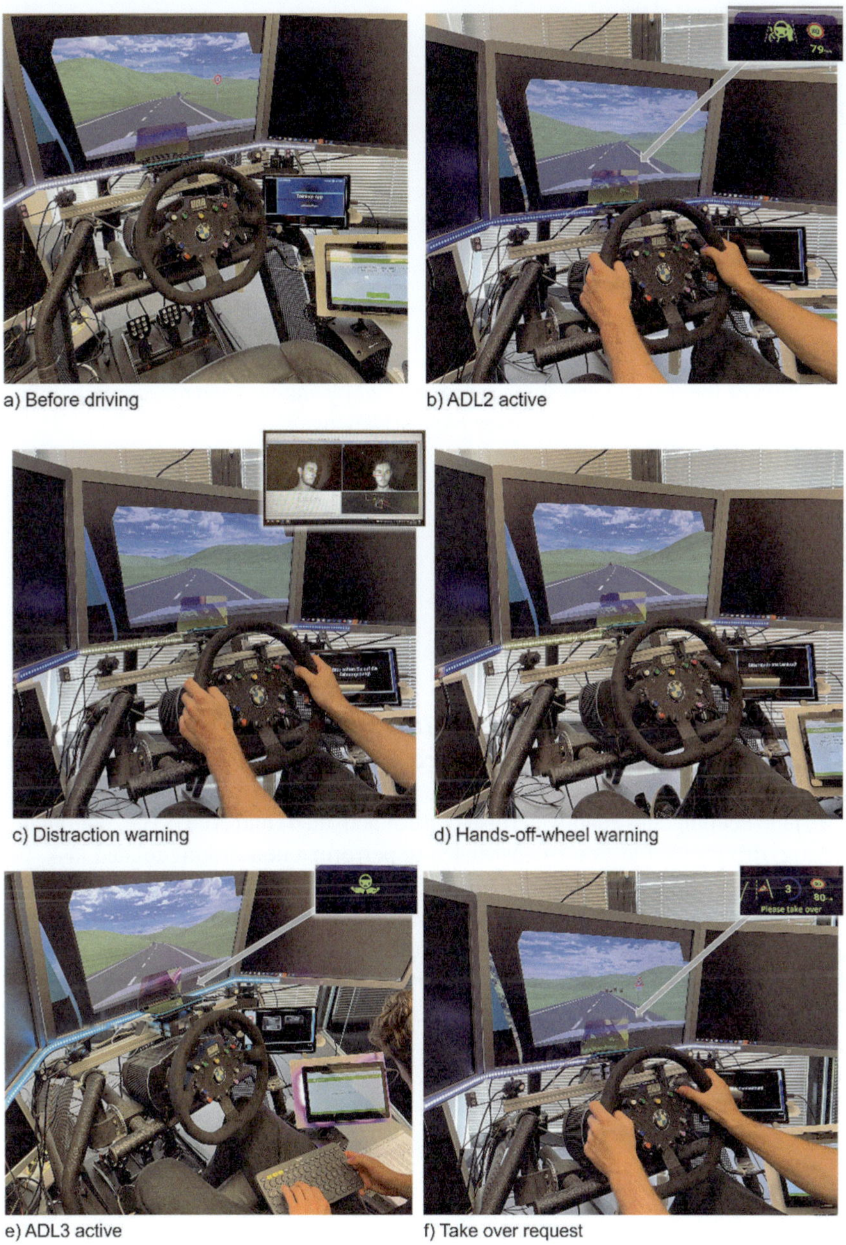

a) Before driving
b) ADL2 active
c) Distraction warning
d) Hands-off-wheel warning
e) ADL3 active
f) Take over request

Fig. 3 Different system states of the HADRIAN integrated fluid HMI

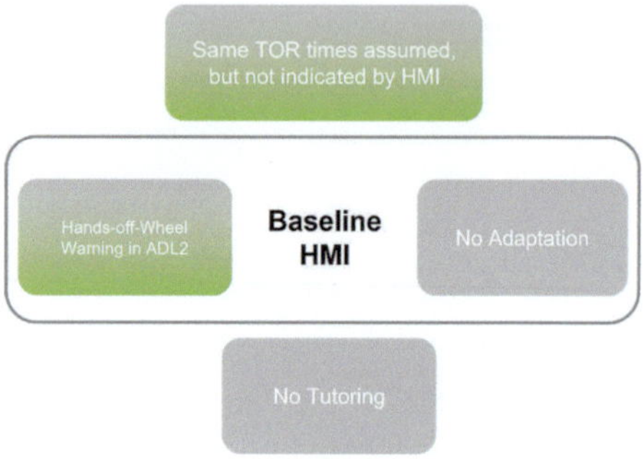

Fig. 4 Key features of the baseline HMI

activation of the respective driving function (i.e., by showing the respective symbols for ADL2 and ADL3), and take-over request messages.

5 Experimental Study

5.1 Study Design

Both HMIs were compared in a driving simulator study. The study was realized as between-subjects design with *kind of HMI* (baseline HMI vs. HADRIAN HMI) as independent variable. Subjects in each group were performing a drive in the simulator, which included manual driving sections as well as sections with ADL2 (i.e., lane keeping and adaptive cruise control activated) and ADL3 (i.e., automated driving) driving. Subjects were instructed to perform a non-driving related task. For that, riddles popped up at a tablet mounted at the center stack in regular intervals, i.e., if the subject completed a riddle and sent the answer, the next riddle would pop up after approximately one minute. The riddles could be answered with a provided Bluetooth keyboard placed aside the driver seat. During the study, driving behavior, eye movements, hands-off-wheel data, as well as subjective measures (acceptance, safety, comfort, usability, trust, etc.) were captured. Table 1 provides an overview of the captured data and how the data was used to trigger the display of information or warnings in the HMI.

5.2 Participants

In total, 43 subjects participated in the study. Four participants needed to be excluded from data analysis due to system malfunction. Table 2 provides an overview of the gender and age distribution, as well as mean technology affinity scores for each group. There were 20 participants in the baseline group, and 19 in the HADRIAN group.

Table 2 Gender, age distribution, and mean technology affinity scores for the two experimental groups

	Baseline (n = 20)	HADRIAN (n = 19)	Total (n = 39)
Gender	10 males 10 females	10 males 9 females	20 males 19 females
mean age	25.2 years (SD = 3.7)	26.9 years (SD = 5.9)	26.1 years (SD = 4.9)
Mean technology affinity score (1–6)	4.68 (SD = 0.85)	4.24 (SD = 1.05)	

Participants were primarily recruited online via social media (Facebook, LinkedIn etc.).

5.3 Procedure

Figure 5 provides an overview of the study procedure. In general, the procedure was similar for both groups. Participants were welcomed and introduced to the study. After signing an informed consent, participants were asked to fill in a pre-questionnaire. After that, the participants were asked to sit down in the simulator mockup and a calibration of the eye-tracking system was performed. The participants received general instructions about driving in the driving simulator and then performed a practice drive for familiarization. After that, the non-driving related task (riddles task) was introduced, and participants were told to answer the riddles as good as possible.

This was followed by the introduction to the respective system. For the baseline HMI, participants were asked to read a written manual ("Operating and safety information"). The print-out provided an overview how the driving assistance and automated driving function work and how to use it (i.e., how availability is indicated, how to activate the respective function, and how to deactivate it in case of a take-over request). Furthermore, general safety hints were provided, i.e., to get an overview

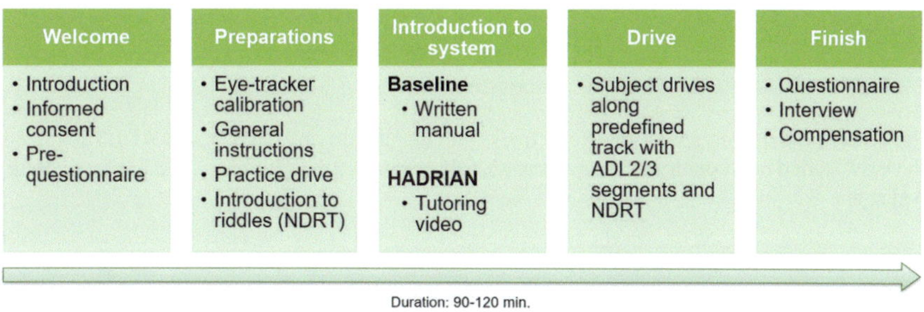

Fig. 5 Study procedure

of the traffic situation before taking over, and it was outlined that the driver has full control and responsibility for driving if driving functions are deactivated.

For the HADRIAN HMI, participants were shown a tutoring video on the HMI tablet explaining the different driving functions, how to use them, and the respective responsibilities of the driver. The video showed how the availability of the respective function is indicated, how to activate it, and what to do in case of a take-over request. At the end of the video, the most important points for using the different driving functions were summarized. It was also pointed out that the tutoring would provide auditory hints during driving and that it would be possible to repeat single video chapters while being in ADL3 mode. Note that the video did not provide information about the distraction or hands-off-wheel warnings to not overwhelm the driver with too much information.

After the introduction to the respective HMI, participants were asked to start driving. Both groups were driving along the predefined study track (see Fig. 6 for a schematic overview) with manual, ADL2, and ADL3 driving segments and eight take-over requests in total. At the end of the drive, the participant had to fill in a questionnaire and a semi-structured interview was conducted. The study duration per participant was approximately 90–120 min and the participant received 20 Euro compensation.

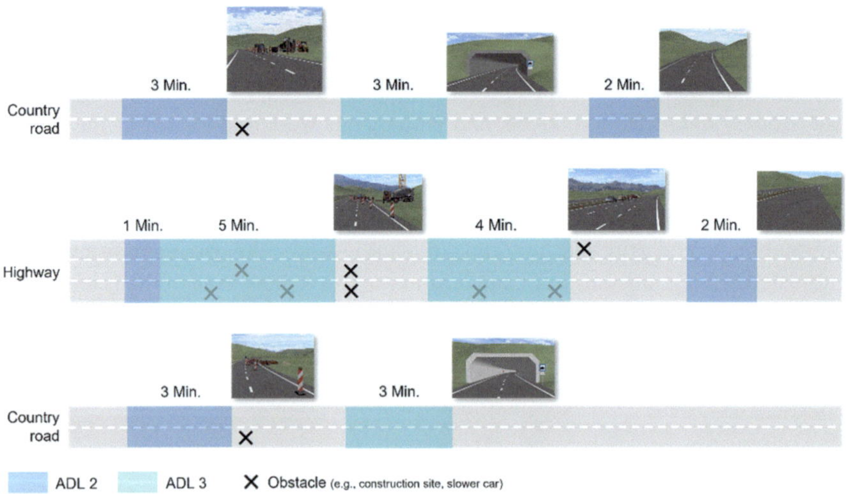

Fig. 6 Schematic overview of the study track with manual driving, ADL2, and ADL3 segments. The track started on a country road (upper row), followed by a 3-lane highway, and finally a country road again

5.4 Technical Setup

The study was set up in a driving simulator consisting of an adjustable driver's seat, a FANATEC Clubsport Racing steering wheel, and three screens facing the driver (see Fig. 2). For the study, the driving simulation was shown on the center screen. The whole study track was created with CarMaker 10.2.1. The track (see Fig. 6 for a schematic overview) started on a country road, followed by a segment on a 3-lane highway and a final country road segment, and had varying speed limits (80, 100, 130 km/h). Four segments of ADL2 and five segments with ADL3 driving were defined and different situations shown in the simulation could trigger a take-over request (construction sites, tunnels, or misleading/missing road markings). There was oncoming traffic on the whole track and traffic on the highway segment. On the highway, the car would perform overtaking maneuvers on its own during ADL3 driving. Buttons on the steering wheel could be used to activate/deactivate the driving functions, and the steering wheel would turn in accordance with the movements of the car when driving functions were active. Driving the whole track took approximately 37 min.

Eye-tracking data was captured with a SmartEye eye-tracking system (60 Hz) with two cameras mounted at the bottom of the center screen. For real-time visual distraction detection, a modified version of the AttenD algorithm (Kircher and Ahlström 2017) was applied. Hands-off-wheel data was captured with a camera mounted at an aluminum profile above the participant filming the steering wheel. Driving data, eye-tracking data, and hands-off-wheel data was synchronized and stored into an Influx database using Virtual Vehicle's Data.Beam,[1] a software platform allowing to connect and synchronize any kind of sensors. As outlined, the simulator and eye tracking system outputs were also used to trigger certain events in the respective HMI, and as basis for the adaptive tutoring provided by the fluid HMI.

The HMI system was served from a Node.js server on a dedicated machine. It received the driving data, eye-tracking data, and the "hands on steering wheel" data from a Python/MediaPipe-based "hands on steering wheel" detection system via separate TCP/IP streams. The server computed all necessary data for the HMI elements and forwarded it to the head-up display via UDP and to a web browser running on an Android tablet via HTTP. The web browser then sent the UI interaction data back to the sever. An Arduino LED controller was connected to the server machine with a USB/Serial connection to receive signals regarding which colors to display on the LED stripes. The server sent all relevant measurement/experiment data to Data.Beam.

Figure 7 depicts the dataflow of the setup on a high-level.

[1] https://www.v2c2.at/databeam.

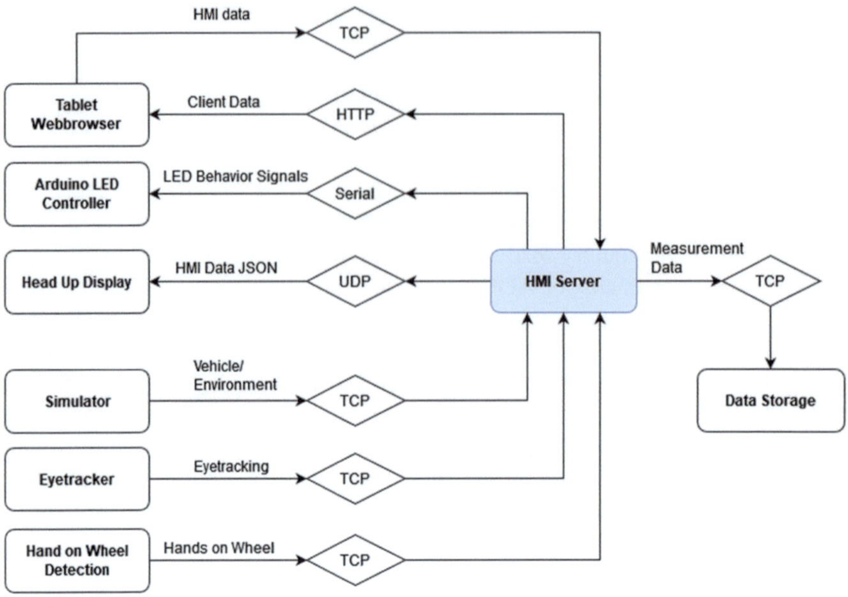

Fig. 7 Data flow of HADRIAN integrated fluid HMI

6 Results

In this section, we present the results of the study, first focusing on subjective measures gained with questionnaires. Then an overview of the objective measures and a conducted safety analysis is provided, followed by a section about the interview results.

6.1 Subjective Measures

Participants answered different questionnaires after completing the drive. Figure 8 provides an overview of the results for the subscales of the UEQ. It is apparent that both HMIs received positive ratings for all scales with the highest ratings for perspicuity, i.e., how easy it was to get familiar with the system. However, we could not find any significant differences between the systems for each of the subscales.

Furthermore, participants had to rate different statements regarding acceptability, usability, trust, safety feeling, and control feeling on a scale from 1 = totally disagree to 7 = totally agree (see Fig. 9). Again, we could not find any significant differences between the ratings for the different systems. While the mean ratings for acceptability, usability, safety feeling, and control feeling are rather high, it is apparent that the ratings for trust are rather medium.

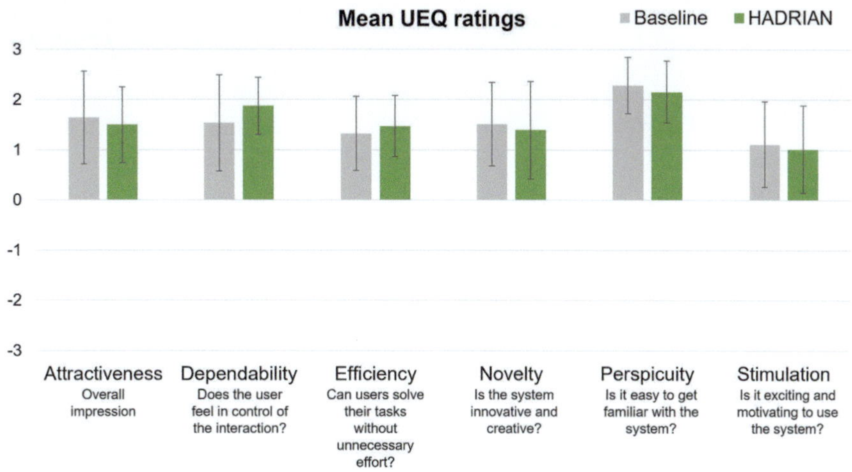

Fig. 8 Mean UEQ ratings for baseline and HADRIAN system

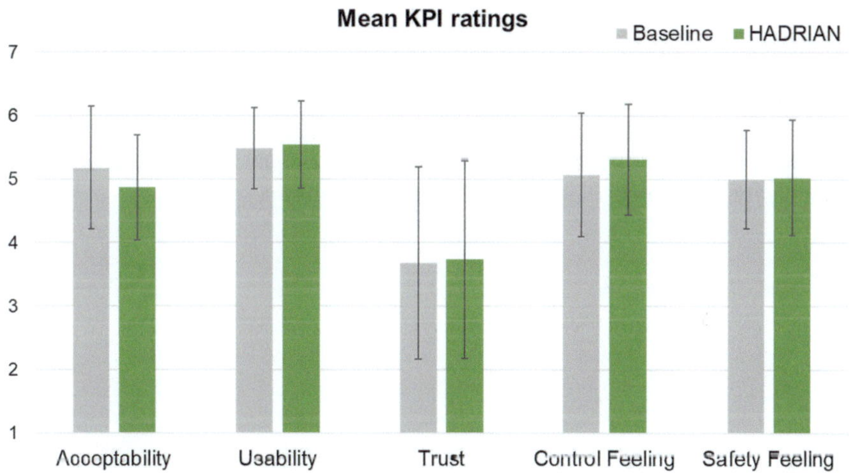

Fig. 9 Mean ratings for acceptability, usability, trust, control feeling, and safety feeling for baseline and HADRIAN system

Furthermore, participants were asked to indicate their demand for the different subscales (mental/physical/temporal demand, performance, effort, frustration) of the NASA-TLX questionnaire. Again, no significant differences between systems could be found (see Fig. 10).

To determine participants' understanding (mental model) of the functionalities of the respective system and their responsibilities, we asked several open questions about their knowledge and understanding. We scored participants' responses based on their correctness. If no answer was given, the score was 0. Otherwise, the maximum

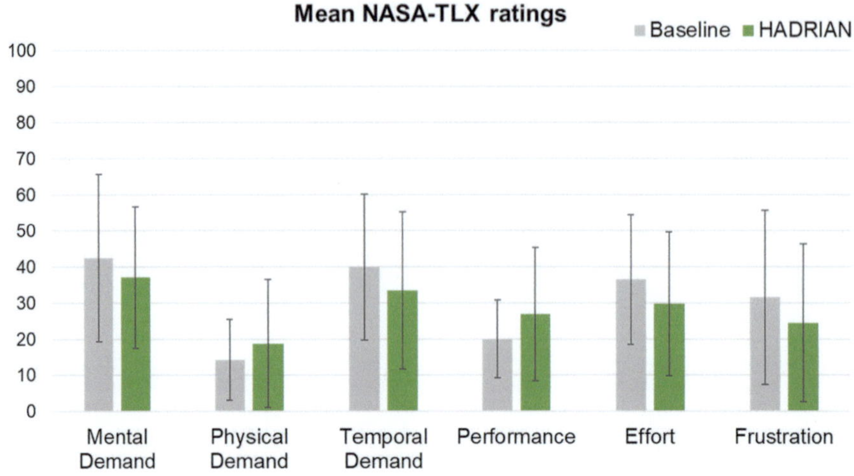

Fig. 10 NASA-TLX ratings for baseline and HADRIAN system

score for a completely correct answer was 1 and was reduced proportionally for incorrect statements. Figure 11 provides an overview of the results.

We found that particularly for questions regarding the tasks and responsibilities of the driver (*What are your tasks when driving automation is active? What should be done as soon as automated driving is no longer available?*) could be answered significantly better by participants of the fluid HMI group.

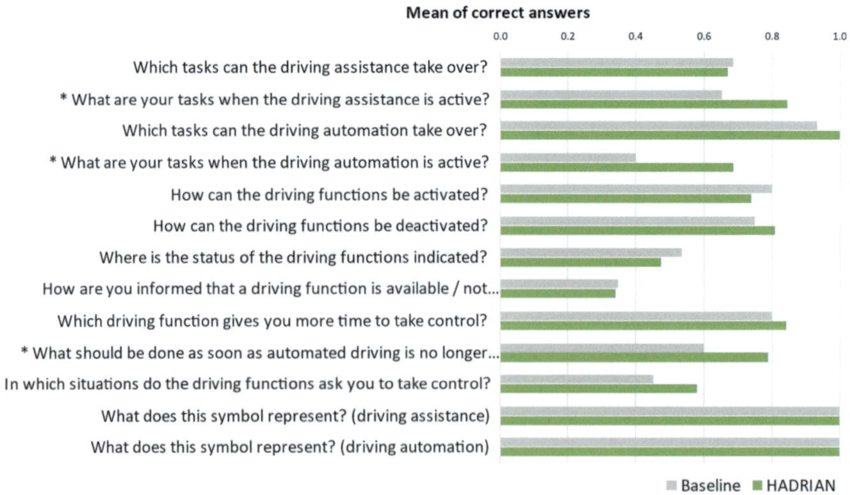

Fig. 11 Mean of correct answers for different knowledge and understanding questions regarding the systems (*$p < 0.05$)

The results also show that both groups could correctly assign the respective symbol for ADL2/ADL3 to the correct function, indicating that the symbol design was intuitive. In the HADRIAN group only, we also asked which LED color represents driving assistance (blue) and which color represents driving automation (cyan). Here, we found a mean score of 0.33, indicating that participants had trouble to assign the colors correctly.

6.2 *Objective Measures*

Participants were asked to perform a secondary task while driving. Hence, we were interested in determining the level of visual distraction at each automated driving level. We calculated the amount of time (in %) during which the gaze of participants was off the road (i.e., off the screen showing the driving simulation) for the different driving levels. Figure 12 provides an overview of the results, which reveal a clear advantage of the fluid HMI over the baseline HMI. The amount of gaze off the road is significantly less for the fluid HMI during manual and ADL2 segments. During manual driving participants of the baseline group had their gaze off the road 9.89% of the time—for the HADRIAN group the amount was 5.46% ($p < 0.05$, student's t-test). Particularly during ADL2 driving, the amount of gaze off the road was substantially lower (10.62%) in the fluid compared to the baseline condition (31.83%, $p < 0.001$).

In contrast, for ADL3, where it was allowed to perform the NDRA, we found that the amount of gaze off the road was significantly higher for the fluid HMI group (65.27%) compared to the baseline (47.53%, $p < 0.001$). This indicates that the fluid

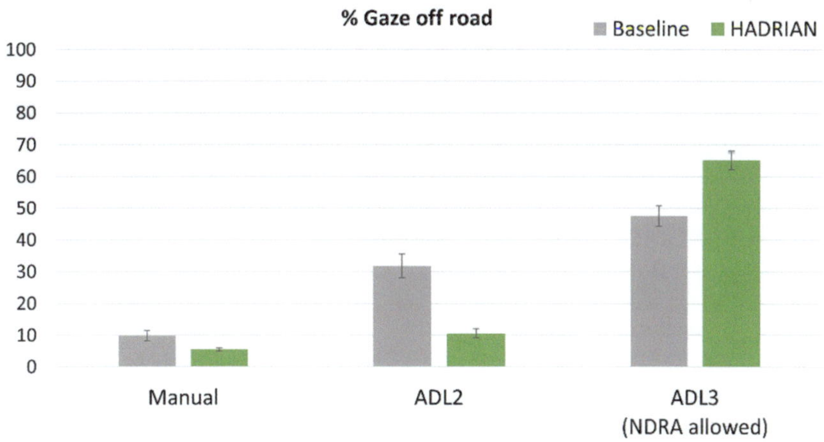

Fig. 12 Gaze off the road in % for the different driving levels and different HMIs

HMI group had a better mode awareness and better understanding of their responsibilities as drivers in the different modes. This is also supported by the participants responses reported above (Fig. 11).

A safety and impact assessment was developed to analyze and evaluate the improvements achieved through the HADRIAN fluid HMI In this study, only the safety part is analyzed, and not the impact part, since the subjective measurements were discussed thoroughly previously. To assess the HADRIAN HMI enhancements, the HADRIAN HMI was compared with the baseline HMI, as in the aforementioned results. The safety assessment methodology was tailored to HADRIAN and was developed using specialized and specific Key Performance Indicators (KPIs). The KPIs were estimated through driving and eye-tracking data that were recorded during the study. The assessment included 9 KPIs for safety and driving performance. The list of the 9 KPIs is included in Table 3, and the methodology for concluding these KPIs as well as the full definition of the indicators are described in Sekadakis et al. (2022).

With the use of these 9 KPIs, a total score was calculated using Data Envelopment Analysis (DEA; Ramanathan 2003) to obtain scores for both the baseline and HADRIAN HMIs for comparison purposes. DEA is the benchmarking method used to evaluate the relative efficiency of each driver based on the KPIs as an input and the distance covered as an output. DEA produced efficiency scores that can be translated into a scoring ranging from 0 to 100%. The extracted efficiency scores are illustrated in a boxplot (Fig. 13). The results revealed that the studied HADRIAN HMI had a statistically significant higher safety score than the baseline conditions, as determined by a Mann Whitney U test ($p < 0.05$). Moreover, the safety score for

Table 3 KPI list for safety and driving performance

KPI ID	KPI	
KPI 1.1	Take over Maneuver safety evaluation	
KPI 1.2	Take over request awareness time	
KPI 1.3	Take over time	
KPI 1.4	Distraction	
KPI 1.5	Conflicts	
KPI 1.6	Automation engagement	• Level 2 • Level 3
KPI 1.7	Time-to-collision	
KPI 1.8	Number of transitions	• AD → Manual • Manual → AD
KPI 1.9	Driving measurements	• Speeding Duration • Speed Over the Limit • Harsh Cornering • Harsh Braking • Harsh Accelerations

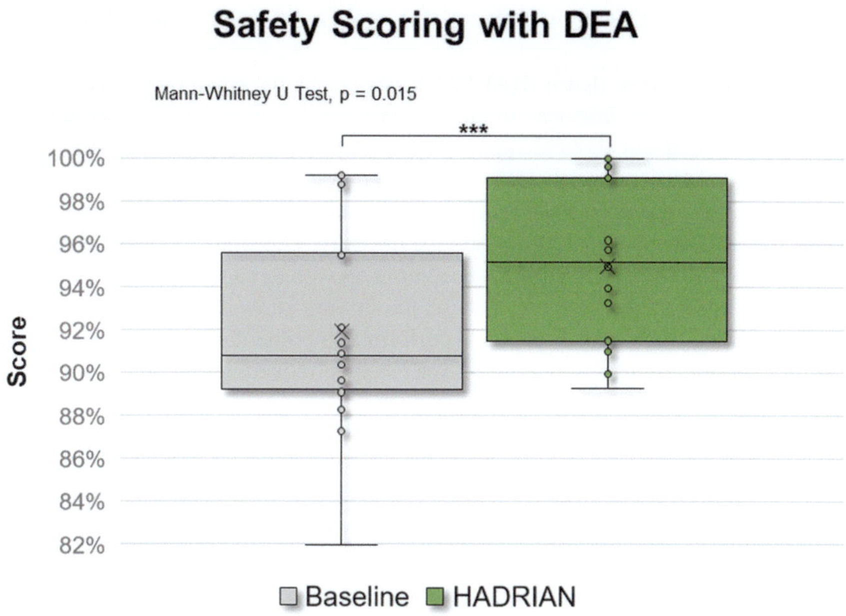

Fig. 13 Safety Scoring using DEA for the different conditions

the HADRIAN HMI had a higher mean, median, and range of scores, indicating a significant increase in safety scoring, and consequently at AD safety level.

The safety enhancement can be attributed to KPIs that were beneficial for safety namely, increased safety score of take over maneuver for the majority of the observations (KPI 1.1, $p = 0.306$, Student's t-test), decreased takeover request awareness time for the majority of the observations (KPI 1.2, $p = 0.373$, Mann–Whitney U test), significantly increased take over time (KPI 1.3, $p = 0.009$, Mann–Whitney U test), significantly limited distraction (KPI 1.4 $p = 0.000$, Mann–Whitney U test), observed with the HADRIAN HMI. The rest of the KPIs had either a negative or a neutral effect on safety. If the rest of the KPIs were more beneficial and positive to safety scoring, a higher safety improvement would have been demonstrated. The full report with the assessment results can be found in deliverable 5.4 of the HADRIAN project (Mörtl et al. 2023).

Incorporating KPIs of Take Over Maneuver Safety Evaluation, Take Over Request Awareness Time, and Take Over Time, it demonstrated the beneficial effect of HADRIAN conditions on takeover performance. Not only the driver is more prepared for a takeover request from automated to manual driving with a longer takeover time but, with the indications from the HADRIAN HMI, the driver takes less time to scan the necessary information from the driving environment to react appropriately prior to take-over and performs a smoother take-over maneuver. The increased safety score should have also benefited from the reduced distraction since the driver is urged by the HMI to stay in the loop with the driving task, as extracted by KPI showing the driver's distraction.

The outcomes show that in HADRIAN conditions the situational awareness of the driver prior to the take-over request is presented higher; participants are well informed and prepared about the upcoming maneuver due to the tutoring before and during the simulation experiment as well as the system informs the drivers for the remaining time until the takeover request. Additionally, during the takeover maneuver phase, the HADRIAN system supports the driver by indicating the level transition, the upcoming obstacles, the current speed limit, as well as haptic cues at the steering wheel. The state of the driver as well as the driving support during the take-over predisposes and supports the drivers to perform a smoother takeover maneuver with less harsh accelerations and braking as well as less frequent speed exceedance.

6.3 Interview Results

For the analysis of the interviews, the interviews were transcribed, categorized, and analyzed using pivot tables in Microsoft Excel. In the following, we report the main findings.

In general, 65% of participants in the baseline condition and 74% in the HADRIAN condition had a positive first impression about the respective system. Particularly the driving functions themselves, i.e., driving assistance and driving automation, were positively outlined in both groups ("*The driving automation was a cool experience because you could do other things while driving*"). In the HADRIAN group, it was also pointed out that the LEDs were nice, and that the tutoring video was "*super and easy to understand*". We also asked, whether the two different driving functions (i.e., driving assistance, automated driving) were clearly distinguishable or whether there had been problems differentiating them. 90% of participants in the baseline condition and 84% in HADRIAN stated that the functions were clearly distinguishable. When being asked how the functions were differentiated, the ADL symbols were mentioned most often in both groups (95% in baseline, 84% in HADRIAN). In the HADRIAN group it was further mentioned that the hands-off-wheel/distraction warning, and the verbal hints while driving were helpful ("*When it was assistance only, there were warnings. That's when I knew at the latest*").

We further asked participants, which of the different system components were most important / most helpful when using the system. In both groups the HUD (80% in baseline, 74% in HADRIAN) and the sounds (55% in baseline, 47% in HADRIAN) were outlined as the most important components. "*The HUD showed everything that was relevant.*" "*Because it was positioned very well, and it showed all information I needed*" "*It was in the field of view all the time*". "*Sounds, because I notice what is happening even when I look away*".

For the HADRIAN system, also other components were considered as useful. Seven participants (37%) outlined the HMI tablet. "*It was helpful to see what the car is doing in automated mode*". "*The time you have still left in automated driving—that was helpful*". Furthermore, the tutoring during and before the drive was experienced as helpful by 37% of participants. "*The voice was very helpful at the beginning to understand what is happening, but it was also good that it went away with time, otherwise it would have been annoying.*" Three participants (16%) considered the LEDs as helpful: "*The LEDs because that's how I recognized whether I had activated a system or not, but I couldn't tell the colors apart.*"

We also asked participants which components they considered less helpful. For the baseline group the HMI tablet was mentioned most often (by 75% of participants), since it did not provide any further relevant information that was not also displayed in the HUD. The rest of participants (25%) stated that nothing was less helpful. For the HADRIAN group, the LEDs were mentioned most often (53%). Here, the main issue was that the colors of the LEDs were not clearly distinguishable for participants and that the LEDs were too obtrusive. "*The colors should be more distinct—don't just use shades of blue.*" "*Design the LEDs more decent and position them better, because they were disruptive.*"

Furthermore, the HMI tablet ("*I least looked at it since everything relevant was on the HUD anyway. There was no added benefit*"; 21%) and steering wheel vibrations ("*The steering wheel vibrations were not very noticeable and can actually be omitted*", 11%) were considered as less helpful by some participants. Five participants (26%) stated that nothing was less helpful.

Finally, we also asked participants in the HADRIAN group what could be improved about the individual system components. Fifteen participants (79%) suggested improvements of the LEDs, primarily to change the colors and make them better distinguishable, to have a more decent/ambient design, and to provide more explanation about the colors. For the HMI tablet, eleven participants (58%) primarily suggested that the tablet should be better positioned, i.e., more in the field of view of the driver, that it could show more information like, e.g., navigation information. As for the steering wheel vibrations, ten participants (53%) stated that they were too rough and should be made smoother, that they were rather distracting/confusing and should be omitted, or that they were not perceived at all.

Regarding the tutoring, six participants (32%) stated to keep the adaptiveness of the tutoring and that it could make sense to incorporate a rewarding system. For the sound, it was mentioned by six participants (32%) that other sounds in the car need to be considered for the sound design and that the sound quality could have been better. For the HUD, it was stated by five participants (26%) that it could also show the remaining time in ADL3 and current speed in ADL3 (note that the HUD would only show the ADL3 symbol when ADL3 was active, see Fig. 3e; everything the car was doing would be displayed on the HMI tablet).

7 Discussion

In general, our results show a discrepancy between subjective and objective measures. While we could not find any significant differences between the baseline HMI and the HADRIAN HMI in the questionnaire ratings, behavioral data and the safety analysis show an advantage of the HADRIAN HMI over the baseline, and it seems that participants indeed had a better mode awareness and safer driving behavior with the HADRIAN HMI.

One important point here is, that the questionnaire ratings were in general rather positive for both systems. Hence, both systems were considered as acceptable and useful. There are different possible reasons for this judgment. As outlined in the final interviews, participants in both groups considered the HUD and the sound as the most helpful components of the respective HMI, while the ADL2/ADL3 symbols served as the most important component to differentiate between driving modes. Hence, components that both HMIs had in common, were considered as most important, which is a probable reason that subjective ratings assimilated accordingly.

Another point is that participants were asked to rate the system as a whole, i.e., "*the whole system with all its components, i.e., the automated driving functions, as well as all displays and hints for communication*". Hence, the driving functions are also a part of the provided system definition, and those functions worked equally well in both conditions. This may be another reason, why subjective ratings were similar. Indeed, the driving functions were also mentioned most often in the final interviews when participants were asked about their first impression about the system, which may be a hint that this influenced the ratings.

The positive impact of the HADRIAN HMI, however, is strongly apparent when it comes to mode awareness and objective measures. Eye-tracking data showed that participants of the HADRIAN group took their eyes off the road significantly less during manual driving, and particularly during ADL2 driving. At the same time, they took their eyes off the road significantly more when it was allowed in ADL3 driving. This indicates that they had a better awareness of what they were allowed to do or not to do in the respective driving mode. As outlined in the final interviews, the tutoring before and during the drive, as well as the distraction/hands-off-wheel warnings can be seen as the main contributing components to this. Furthermore, the safety analysis showed a significant benefit of the HADRIAN HMI over the baseline.

These results were also reflected in the knowledge of participants about their responsibilities in the different driving modes and what to do in case of a take-over request. Here, participants of the HADRIAN group had a significantly better understanding than the baseline group. However, the results also show that not all questions could be answered better by the HADRIAN group, indicating that there is still room for improvement.

The aim of the study was also to determine what can be learned for the design of the integrated HADRIAN HMI. Here our results suggest keeping the HUD, ADL symbols, and sounds as is, since those components were considered as most helpful. Also, the design of the ADL symbols proved to be intuitive and clearly assignable to the respective mode for all participants. As for the information displayed on the HUD, it is an open question, though, whether the depiction of obstacles/closed lanes ahead during a take-over request really provided a benefit. Participants were not asked about this feature precisely, however, also did not outline it when talking about the HUD component in the final interview. This may be an indicator that this feature was not experienced as particularly helpful or outstanding.

As for the distraction and hands-off-wheel warnings, our results showed a clear benefit in terms of safer driving behavior. Hence, these warnings can be considered as important part of the fluid HMI.

The same applies to the tutoring component. Tutoring turned out to be helpful in conveying driver responsibilities, and particularly the adaptiveness of the tutoring (to the respective situation, the take-over behavior, and the timeline of interaction) was appreciated by participants. A crucial point here is, that the tutoring really needs to be tailored as accurate as possible to the driver behavior and needs. If there is a mismatch or upcoming feeling of paternalism, the acceptance of tutoring could easily drop. In our study, it was mentioned by several participants that especially the adaptiveness made tutoring a good feature and should be kept.

For the ambient light, our results show that participants had problems differentiating the colors blue and cyan and correctly associating them with the respective driving mode. Here it was mentioned most often that the colors should be more distinct, and that the design should be more decent in general. One conclusion out of this is that in our study the different color coding for the different automation modes obviously did not support better mode awareness (at least not on a conscious level). Hence, the question is whether it would be more beneficial to just indicate that automation (no matter if it's ADL2 or ADL3) is activated by one color and in addition highlight an allowed NDRA in automated driving levels like ADL3 where it is allowed. This would need further investigation, though. As regards the more decent design, it must be outlined that participants were sitting quite close to the screen and LEDs in our simulator setup, and it is to be expected that in a real vehicle this would be less of an issue. Otherwise, LEDs with a less minimum brightness or a translucent cover could help to solve this issue.

As for the HMI tablet, the provided time information (remaining time in ADL3) and display of the overtaking maneuvers the car performed on its own in ADL3 were considered as helpful. However, it was outlined that the tablet could be better positioned (more in the line of sight of the driver) and could provide additional navigational information. The latter point was not yet realized in this study; however, it is generally pursued in HADRIAN to be able to show the driver upcoming segments of ADL2 and ADL3 driving along with navigational information, i.e., a map indicating such driving segments.

As for the steering wheel vibrations or haptic icons, their added value could not be clearly determined in this study. We believe, however, that this probably also was an implementation issue since we used a different steering wheel with different capability to vibrate compared to the project partner's original steering wheel. This may also be the reason why the haptic icons could not be perceived as well by our study participants. Further investigations need to be done.

8 Conclusion

Our results show an advantage of the HADRIAN integrated fluid HMI over the baseline HMI in terms of behavioral data and with regard to mode awareness and safety, while usability and acceptability were experienced similarly high for both HMIs. Furthermore, the study provided us with valuable insights on how to further improve and adjust the fluid HMI. These insights were taken into consideration for the implementation of the HMI in the field demo (see chapter "Results of two Demonstrations of Holistic Solutions for Automated Vehicles to Increase Usefulness and Safety").

Acknowledgements The development of the HMI and realization of the study was a real joint effort. Therefore, we want to sincerely thank the following people for their contribution: Erika Santuccio, Manuela Prior, Christian Groß, Udo Grossschedl, and Peter Sammer from Virtual Vehicle Research GmbH (Austria); Magdalena Gärtner, Vivien Wallner, and Jakub Sypniewski from Center for Human-Computer Interaction (Austria); Sergio Enrique Diaz Briceño and Mauricio Marcano Sandoval from Tecnalia (Spain), as well as Christos Katrakazas, Marianthi Kallidoni, and George Yannis from NTUA (Greece).

HADRIAN has received funding from the European Union's Horizon 2020 research and innovation programme under grant agreement No 875597. This document reflects only the author's view, the Climate Innovation and Networks Executive Agency (CINEA) is not responsible for any use that may be made of the information it contains. The publication was written at Virtual Vehicle Research GmbH in Graz and partially funded within the COMET K2 Competence Centers for Excellent Technologies from the Austrian Federal Ministry for Climate Action (BMK), the Austrian Federal Ministry for Labour and Economy (BMAW), the Province of Styria (Dept. 12) and the Styrian Business Promotion Agency (SFG). The Austrian Research Promotion Agency (FFG) has been authorised for the programme management.

References

Boelhouwer A, van den Beukel AP, van der Voort MC, Martens MH (2019) Should I take over? Does system knowledge help drivers in making take-over decisions while driving a partially automated car? Transp Res F: Traffic Psychol Behav 60:669–684. https://doi.org/10.1016/j.trf.2018.11.016

Boelhouwer A, van den Beukel AP, van der Voort MC, Verwey WB, Martens MH (2020b) Supporting drivers of partially automated cars through an adaptive digital in-car tutor. Information 11(4):185. https://doi.org/10.3390/info11040185

Boelhouwer A, van den Beukel AP, van der Voort MC, Hottentot C, de Wit RQ, Martens MH (2020a) How are car buyers and car sellers currently informed about ADAS? An investigation among drivers and car sellers in the Netherlands. Transp Res Interdiscip Perspect 4. https://doi.org/10.1016/j.trip.2020.100103

DeGuzman CA, Donmez B (2021) Drivers still have limited knowledge about adaptive cruise control even when they own the system. Transp Res Record: J Transp Res Board 2675(10):328–339. https://doi.org/10.1177/03611981211011482

Dönmez Özkan Y, Mirnig AG, Meschtscherjakov A, Demir C, Tscheligi M (2021) Mode awareness interfaces in automated vehicles, robotics, and aviation: a literature review. In: 13th international conference on automotive user interfaces and interactive vehicular applications. AutomotiveUI '21, Association for Computing Machinery, New York, NY, USA, pp 147–158. https://doi.org/10.1145/3409118.3475125

Ebinger N, Trösterer S, Neuhuber N, Mörtl P (2023) Conceptualisation and evaluation of adaptive driver tutoring for conditional driving automation. In: de Waard D, Hagemann V, Onnasch L, Toffetti A, Coelho D, Botzer A, de Angelis M, Brookhuis K, Fairclough S (2023) Proceedings of the human factors and ergonomics society europe chapter 2023 annual conference. ISSN 2333–4959 (online). Available from http://hfes-europe.org

Ebnali M, Lamb R, Fathi R, Hulme K (2021) Virtual reality tour for first-time users of highly automated cars: comparing the effects of virtual environments with different levels of interaction fidelity. Appl Ergon 90:103226. https://doi.org/10.1016/j.apergo.2020.103226

Forster Y, Hergeth S, Naujoks F, Krems J, Keinath A (2019) User education in automated driving: owner's manual and interactive tutorial support mental model formation and human-automation interaction. Information 10(4):143. https://doi.org/10.3390/info10040143

Franke T, Attig C, Wessel D (2019) A personal resource for technology interaction: development and validation of the affinity for technology interaction (ATI) scale. Int J Hum-Comput Interact 35(6):456–467. https://doi.org/10.1080/10447318.2018.1456150

Hart SG, Staveland LE (1988) Development of NASA-TLX (task load index): results of empirical and theoretical research. In: Advances in psychology, vol 52. North-Holland, pp 139–183. https://humansystems.arc.nasa.gov/groups/tlx/downloads/TLXScale.pdf

Hecht T, Weng S, Kick LF, Bengler K (2022) How users of automated vehicles benefit from predictive ambient light displays. Appl Ergon 103:103762

International Organization for Standardization (ISO) (2011) ISO 3864–4:2011 graphical symbols—safety colours and safety signs—part 4: colorimetric and photometric properties of safety sign materials. https://standards.iteh.ai/catalog/standards/sist/775b486c-7817-4a6a-a7cb-d2b2cc661376/iso-3864-4-2011

Kircher K, Ahlström C (2017) The driver distraction detection algorithm AttenD. In: Driver distraction and inattention. CRC Press, pp 327–348

Koo J, Shin D, Steinert M, Leifer L (2016) Understanding driver responses to voice alerts of autonomous car operations. Int J Veh Des 70(4):377–392

Laugwitz B, Held T, Schrepp M (2008) Construction and evaluation of a user experience questionnaire. In: HCI and usability for education and work: 4th symposium of the workgroup human-computer interaction and usability engineering of the Austrian computer society, USAB 2008, Graz, Austria, November 20–21, 2008. Proceedings 4. Springer Berlin Heidelberg, pp 63–76

Löcken A, Heuten W, Boll S (2016) Autoambicar: using ambient light to inform drivers about intentions of their automated cars. In: Adjunct proceedings of the 8th international conference on automotive user interfaces and interactive vehicular applications, pp 57–62

Marx C, Ebinger N, Santuccio E, Mörtl P (2022) Bringing the driver back in-the-loop: usefulness of letting the driver know the duration of an automated drive and its impact on takeover performance. In: 13th international conference on applied human factors and ergonomics (AHFE 2022), New York, USA. https://doi.org/10.54941/ahfe1002475

Mirnig AG, Gärtner M, Wallner V, Demir C, Dönmez Özkan J, Sypniewski J, Meschtscherjakov A (2023) Enlightening mode awareness: guiding drivers in the transition between manual and automated driving modes via ambient light. Personal and ubiquitous computing—theme issue on engaging with automation: understanding and designing for operation, appropriation, and behavior change. Springer. (in production)

Mörtl P, Gross C, Sekadakis M, Kallidoni M, Katrakazas C (2023) Overall effectiveness assessment tool and final effectiveness assessment. Final version as of 30.05.2023 of deliverable D5.4 of the HORIZON 2020 project HADRIAN. EC Grant agreement no: 875597. https://hadrianproject.eu/results-2/

National Electrical Manufacturers Association (NEMA) (2022) ANSI Z535.1–2022 American national standard for safety colors. https://www.nema.org/standards/view/american-national-standard-for-safety-colors

Neuhuber N, Ebinger N, Pretto P, Kubicek B (2022) How am i supposed to know? Conceptualization and first evaluation of a driver tutoring system for automated driving. Human factors and ergonomics society Europe chapter annual meeting. HFES Europpe 2022. https://www.hfes-europe.org/wp-content/uploads/2022/05/Neuhuber2022.pdf

Neuhuber N, Ebinger N, Kubicek B (submitted). What is trust, anyway? Towards the design of a practical trust scale. Theor Issues Ergon Sci

Osswald S, Wurhofer D, Trösterer S, Beck E, Tscheligi M (2012) Predicting information technology usage in the car: towards a car technology acceptance model. In: Proceedings of the 4th international conference on automotive user interfaces and interactive vehicular applications, pp 51–58

Payre W, Cestac J, Dang N-T, Vienne F, Delhomme P (2017) Impact of training and in-vehicle task performance on manual control recovery in an automated car. Transp Res F: Traffic Psychol Behav 46:216–227. https://doi.org/10.1016/j.trf.2017.02.001

Pretto P, Mörtl P, Neuhuber N (2020) Fluid interface concept for automated driving. In: HCI in mobility, transport, and automotive systems. Automated driving and in-vehicle experience design: second international conference, MobiTAS 2020, held as part of the 22nd HCI international conference, HCII 2020, Copenhagen, Denmark, July 19–24, 2020, Proceedings, Part I 22. Springer International Publishing, pp 114–130

Ramanathan R (2003) An introduction to data envelopment analysis: a tool for performance measurement. SAGE

Revell KM, Brown JW, Richardson J, Kim J, Stanton NA (2021) How was it for you? comparing how different levels of multimodal situation awareness feedback are experienced by human agents during transfer of control of the driving task in a semi-autonomous vehicle. In: Designing interaction and interfaces for automated vehicles. CRC Press, pp 101–113

Rukonić L, Mwange M-A, Kieffer S (2022) Teaching drivers about ADAS using spoken dialogue: a wizard of Oz study. In: Proceedings of the 17th international joint conference on computer vision, imaging and computer graphics theory and applications. SCITEPRESS—Science and Technology Publications, pp. 88–98. https://doi.org/10.5220/0010913900003124

SAE International (2021) Taxomony and definitions for terms related to driving automation systems for on-road motor vehicles (J3016)

Sarabia J, Diaz S, Marcano M, Zubizarreta A, Pérez Rastelli J (2022) Haptic steering wheel for enhanced driving: an assessment in terms of safety and user experience. In: Adjunct proceedings of the 14th international conference on automotive user interfaces and interactive vehicular applications, pp 219–221

Sekadakis M, Katrakazas C, Clement P, Prueggler A, Yannis G (2022) Safety and impact assessment for seamless interactions through human-machine interfaces: indicators and practical considerations. Transp Res Arena (TRA) 2022, Lisbon, Portugal, 14–17 Nov 2022

Selcon SJ, Taylor RM (1990) Evaluation of the situation awareness rating technique (SART) as a tool for aircrew systems design. In: AGARD conference proceedings No. 478 (pp. 5/1–5/6). ISBN 92-835-0554-9

Strand N, Nilsson J, Karlsson I, Nilsson L (2011) Exploring end-user experiences: self-perceived notions on use of adaptive cruise control systems. IET Intel Transp Syst 5(2):134. https://doi.org/10.1049/iet-its.2010.0116

Wang J, Wang W, Hansen P, Li Y, You F (2020) The situation awareness and usability research of different hud hmi design in driving while using adaptive cruise control. In: International conference on human-computer interaction. Springer, pp 236–248

Wintersberger P, Dmitrenko D, Schartmüller C, Frison AK, Maggioni E, Obrist M, Riener A (2019) S(c)entinel: monitoring automated vehicles with olfactory reliability displays. In: Proceedings of the 24th international conference on intelligent user interfaces, pp 538–546

Automated Driving Vehicle Functionality as Guardian Angel

Joseba Sarabia, Sergio Diaz, Mauricio Marcano, Alexander Mirnig, and Bharat Krishna Venkitachalam

Abstract The concept of the Guardian Angel system represents a pivotal advancement in vehicular safety, with a focus on enhancing the driving experience for individuals with diminished driving skills, particularly elderly drivers seeking to retain their mobility. This system functions as an unobtrusive co-pilot, intervening only when necessary, and empowering drivers to maintain control while ensuring their safety. By actively monitoring both the external environment and the interior of the vehicle, the Guardian Angel system adeptly identifies potential hazards and triggers interventions in response to imminent collisions, road departures, or internal factors such as driver distraction or drowsiness. Through a comprehensive array of Human–Machine Interfaces (HMIs), the Guardian Angel system communicates critical information to the driver, enhancing situational awareness and facilitating seamless cooperation between humans and machine. The holistic design ensures that the system operates unobtrusively in the background, engaging only in safety–critical situations and providing clear explanations for its interventions. This paper presents a detailed exposition of the Guardian Angel system's architecture, its controller design, and the diverse range of HMIs employed to relay information to the driver. The focus here lies in articulating the system's conceptual foundation, design principles, and the potential it holds for transforming the driving experience into a safer and more empowering endeavor for drivers of varying skills.

Keywords Shared control · Vehicular safety · Elderly drivers · Human Machine Interfaces (HMIs)

J. Sarabia (✉) · S. Diaz · M. Marcano
Tecnalia, Basque Research and Technology Alliance, Derio, Biscay, Spain
e-mail: joseba.sarabia@tecnalia.com

A. Mirnig · B. K. Venkitachalam
Center for Human-Computer Interaction, University of Salzburg, Salzburg, Austria

P. Moertl and B. Brandstaetter (eds.), *Shaping Automated Driving to Achieve Societal Mobility Needs*, Lecture Notes in Mobility,
https://doi.org/10.1007/978-3-031-52550-6_4

1 Introduction

Automated driving technologies are promising, but still require an extended investment in time and effort to achieve a mature state to achieve the SAE levels 4 and 5, where the vehicle becomes responsible of the driving task. In these levels, the driver gets replaced instead of collaborating with him/her. These technologies are meant to improve safety in roads, among others. However, this is a particularly interesting paradox, as a properly trained human driver is proven to be currently the best safety system of the vehicle, and it is expected that it will take long until such safety level is achieved by the automated system.

With this regard, the current study focuses on enhancing the capabilities of the drivers rather than replacing them. The investigated system is the Guardian Angel. It provides a safety envelope that seeks protecting the driver from the hazards surrounding the vehicle, intervening when the vehicle is detected to be in a non-safe state.

Regarding the SAE levels, the Guardian Angel does not exactly fit in the description of any of them. The responsibility of the driving task shifts between the driver and the automated system, depending on the hazards presented. Thus, in terms of driving technical capabilities, the GA requires SAE level 4 technology, so that it can perform safety–critical maneuvers. However, it supports manual driving, like less automated levels (1 to 3).

About the hazards detected by the GA, they can be divided in two groups, external hazards and internal hazards. External hazards group other road users, obstacles or complicated road conditions (Sarabia et al. 2023). Internal hazards, on the contrary, refer to the hazards caused by the driver itself. This can be unsafe behaviors of the driver, distractions, fatigued driving or even diminished driving capabilities, like the case of elderly drivers.

2 Target Driver Profile

The Guardian Angel is conceived with the idea of supporting elderly people in their daily driving routines.

As individuals age, their physical and cognitive abilities may decline, which can impact their ability to drive safely (see Chap. 2). Higher reaction times, for example, may make it more difficult for older drivers to respond to unexpected situations on the road, such as a sudden stop by another vehicle (Makishita and Matsunaga 2008). Additionally, older adults may have less awareness of their surroundings, which can make it more difficult to identify and respond to potential hazards. They may also have difficulty with multi-tasking and may have trouble keeping track of multiple cars and other traffic around them (Mack et al. 2022).

Reduced capacity to control the vehicle is another concern, as older adults may have difficulty with fine motor skills, which can make it harder to steer, brake, and accelerate. They may also have difficulty with spatial awareness and judgement, which can make it harder for them to navigate through busy roads and complex traffic junctions. Additionally, older adults may experience age-related declines in vision, which can reduce visibility, making it more difficult to see other vehicles, pedestrians, and other obstacles on the road (Payyanadan et al. 2018). They may also have trouble with glare and night vision, making it harder to drive in certain lighting conditions.

3 Guardian Angel Description

3.1 *Concept*

The Guardian Angel is intended to ensure the safety of the manual driving of a driver. A typical user of the system is a driver with diminished driving skills. It is primarily targeted towards helping elderly drivers remain mobile after losing some of their driving license privileges (see Chap. 2). Yet, its functionality will also benefit any other user, as drivers may overlook critical unexpected events or respond too slowly under quickly developing safety critical situations.

The tasks of the driver stay the same as during manual driving without Guardian Angel. However, during the course of a drive, the vehicle intervenes the driver's control manoeuvres that seem unsafe under a principle of minimum intervention. At all times the driver should stay in control, thus improving user acceptance, especially in elder drivers that enjoy their independence and reinforce their self-confidence.

The vehicle monitors the surroundings and the interior of the vehicle to identify unsafe situations and define the proper action in a fluid manner. Thus, the system interventions are gradually triggered depending upon external risks, such as an impending collision or road/lane departure event, and upon internal unsafe conditions such as a distracted/sleepy driver.

As long as there is not an unsafe condition, the system will not intervene. Thus, the driver should feel comfortable and should not feel prompted to deactivate the system (de Winter et al. 2022). Only if a safety critical situation emerges that the driver seems not to be handling in a timely manner, the vehicle intervenes and corrects the control actions by guiding or, if needed, overriding the lateral control. The vehicle also provides an indication of why the manoeuvre has happened and what purpose it had.

3.2 *Functionality*

The main feature in the Guardian Angel system is a haptic steering wheel operating under a shared control scheme (Winter et al. 2022). A safety envelope is defined

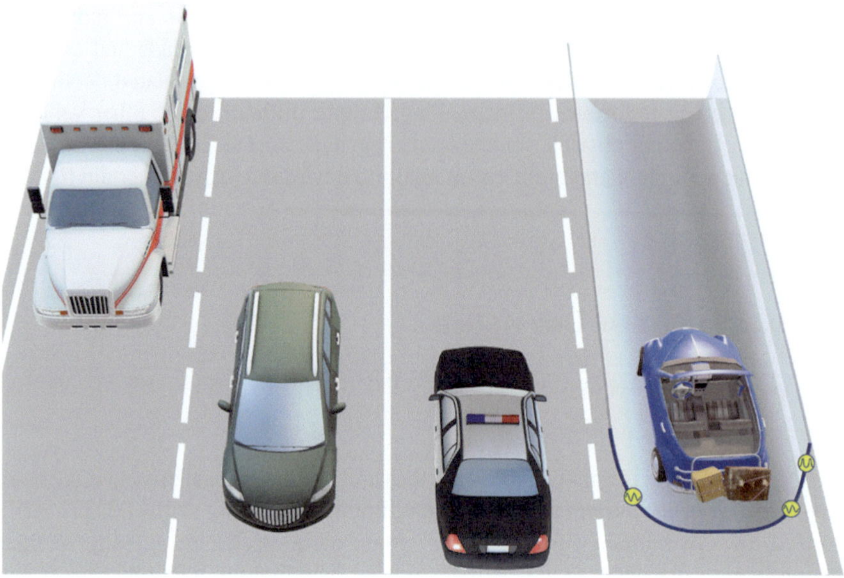

Fig. 1 Safety envelope

as the region where the vehicle stays safe in the road depending on the road and the obstacles, and the driver is free to drive within it (see Fig. 1). If the vehicle approaches the limit of the safety envelope, the system applies torque to the steering wheel in a gradual manner to keep it inside. The higher the crash-risk, the more authority is given to the system, allowing for more torque to be applied (Flemish et al. 2003). Thus, as the situation develops, if a maneuver is deemed to lead to an unsafe condition, the driver will start feeling a gentle guiding torque towards safety, which will grow in intensity if the risk keeps increasing and could override the driver if finally needed.

In addition to the steering support, the system will provide complementary information to increase driver's awareness and understanding of the system actions. Ambient lighting on the windshield and steering wheel (see section Ambient Visual HMI), along with auditive and haptic warnings messages, informs the driver of the source of danger and the steering corrections needed. Lights in the lateral sides of the windshield warn the driver of potential lane departure situations, while the central light in the windshield catches driver attention when distracted. Furthermore, an adaptive tutoring system (Chap. 5, Sect. 3.1.1.5) explains the driver the interventions to improve acceptance and comfort. This system provides real time audio messages to the driver providing extra information about the interventions of the automated system.

The maximum authority that the system can reach is adapted according to the driver's state (see Fig. 2). In that way, a distracted driver will receive a higher level of assistance than an attentive driver. That is, its safety envelope would be smaller and

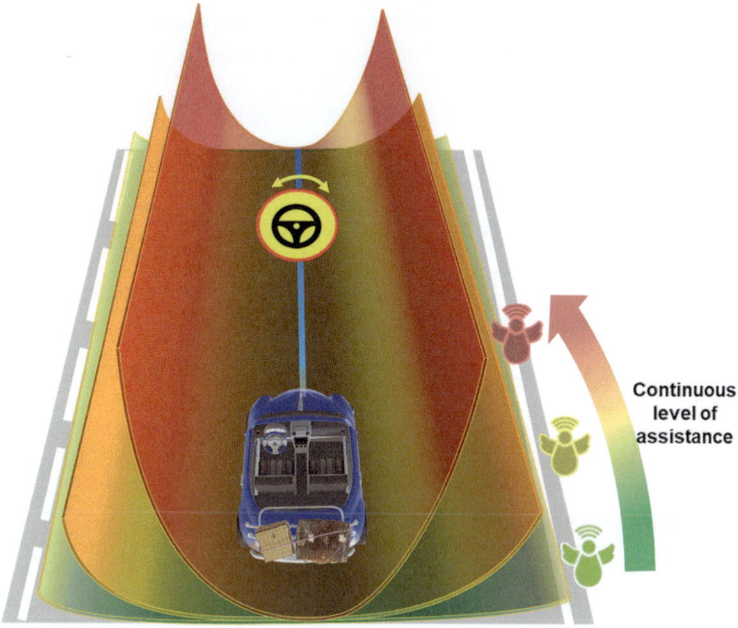

Fig. 2 Continuously varying level of assistance

the supporting torques would be stronger. As a result, a distracted (or sleepy/tired/underperforming) driver will receive more support than one that is more fit-to-drive. This support, however, is only applied for a limited period of time to avoid misuse or abuse. If the unfit driver condition persists, escalating warnings are executed to regain driver's attention and, if it is not recovered, the vehicle stops.

3.3 Collaborative Behavior

The Guardian Angel requires a collaborative behavior between the driver and the automation. However, putting two responsible agents in front of the driving task may bring discrepancies in the decision making. For that reason, the collaboration strategy of the Guardian Angel has two sides.

From the one side, there is the shared control between the driver and the automation. To prevent decision mismatch between the driver and the automation, priority is given to the driver by default. If the driver is showing a perceivably appropriate driving state, the driver is allowed to move the vehicle freely inside the lane. However, if the DMS detects an unsafe driver state, the automation takes over control avoiding the vehicle to depart from the lane (Marcano et al. 2021).

From the other side, the Guardian Angel actively communicates with the driver to inform about its interventions. Also, in case of misuse, it will alert the driver to stop the misuse.

These interactions are managed through the following components:

1. The Guardian Angel Controller that handles the lateral control of the vehicle, depending on the driver state.
2. The Guardian Angel Driver Monitoring System (DMS) that captures in real time the state of the driver.
3. The Guardian Angel HMI, that informs and alerts the driver about the detected external and internal hazards.

In the following section, an extended explanation of the shared control system and the HMI is shown.

4 Guardian Angel Controller

4.1 Shared Control

One of the main challenges in shared control for automated driving is determining the appropriate level of control to allocate to each agent in a given situation. This requires the development of algorithms that can adapt to the changing needs and abilities of the human driver (Marcano et al. 2020). Additionally, shared control systems for automated driving must be designed in a way that is intuitive and easy for the human driver to use, and that allows the driver to maintain a sense of control and situational awareness (Sarabia et al. 2022). Overall, shared control has the potential to improve the performance and safety of automated driving systems by combining the strengths of both humans and computers.

The typical block diagram of the automated vehicle is based on information coming solely from the environment. For shared control, however, the driver must also be monitored, including additional components to support the interaction between the driver and the automation (Marcano et al. 1268).

The shared control module arbitrates between the human driver and the automation. This module receives inputs from both the human driver, through the DMS, and from the automated system. The Adaptive shared control module then determines the appropriate level of control to allocate to each agent in a given situation and generates control signals to be sent to the vehicle's actuators (Fig. 3).

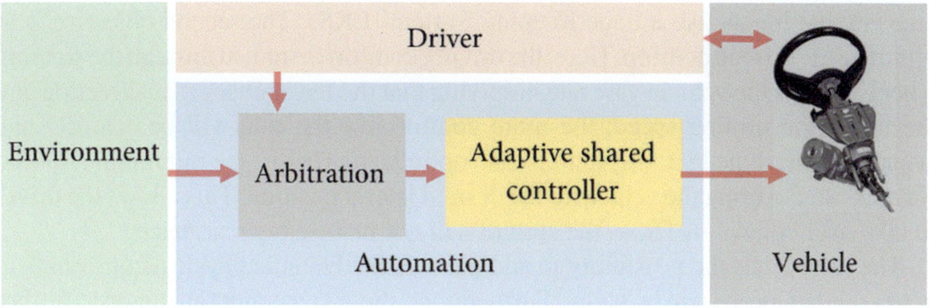

Fig. 3 Block diagram of the Guardian Angel Shared control

4.2 Arbitration

The approach proposed for the Guardian Angel involves both the driver and the automation being in the loop simultaneously. When the driver is fit to drive, the system prioritizes the control actions of the driver. The system is still working behind, supervising for safety hazards. In case of an event compromising safety, the automation performs the required safety maneuver. However, when the driver shows patterns of not being fit to drive, like being fatigued or distracted, or driving unproperly, the automation will override him/her.

4.3 Driver

The driver is responsible for the drive like in a manual driving condition during manual operation. The Guardian Angel does not allow the driver to get involved into secondary tasks, as it is not replacing the driver, but supporting him/her. It is important to mention that the Guardian Angel has the capability to override the driver when safety is compromised, making the automation responsible for the drive during this period of time.

4.4 Adaptive Shared Controller

The control of the Guardian Angel is a torque-based lateral controller based on non-linear Model Predictive Control (nMPC). The nMPC allows to introduce a non-linear mathematical model of the vehicle that more accurately represents the system's behavior (Sarabia et al. 2023).

The chosen optimization problem is focused on three main objectives. First, to minimize the lateral error of the vehicle position with respect to the defined trajectory. In a normal driving situation, the defined trajectory is centered in the middle of the lane, so that it acts like a Lane Keeping System (LKS). The second objective is to optimize the driving comfort. Here, the driving comfort is studied through the steering wheel rate and the vehicle yaw rate, implying that the fewer changes in direction and the lesser the turning speed, the more comfortable the ride will be. Thirdly, the steering effort done by the system is also optimized. That is, the minimum required force is used to bring the vehicle to the desired lateral position. This allows the driver to take over control easier, as the system will not make a big resistance.

The nMPC has the possibility to add the vehicle dynamic physical constraints to better predict the system behavior. Furthermore, these constraints are tuned to further restrict the motion of the vehicle. With that regard, three variables are constrained in the Guardian Angel: the yaw rate, the torque and torque rate exerted in the steering wheel. The yaw rate helps avoiding the vehicle drift, and the torque and the torque rate ensure that it is always possible for the driver to resume control.

As mentioned before, the arbitration block evaluates what level of assistance the driver requires. The adapted variable is the maximum steering torque. The constraint of the maximum steering torque is shifted in terms of the safety hazards perceived. The hazards are obtained through 3 different sources:

- Other vehicles, VRU or road obstacles
- Lateral position of the vehicle in the lane
- Driver fitness to drive

A fuzzy logic controller is used to provide a single output from these three inputs. Fuzzy logic allows to have a comprehensive control solution with intuitive tuning and an easy implementation for multivariable systems like this one. Compared to AI, it allows the designer to have a more insightful vision of the control, rather than a black box model. Each of these variables is defined by triangular membership functions defined by rules (Marcano et al. 2021).

5 Guardian Angel HMI

In the pursuit of seamless driver-system collaboration, the Guardian Angel system employs a diverse range of Human–Machine Interfaces (HMIs) to establish effective communication channels between the vehicle and the driver. These interfaces are thoughtfully designed to empower drivers, particularly those with diminished driving skills, by providing real-time information and guidance while maintaining a balance between intervention and driver autonomy. This section shows all the HMIs that were installed in the vehicle for the application of the Guardian Angel and delves into the description of 2 HMIs: The Ambient Visual HMI, based on LEDs embedded in the windshield and in the steering wheel, and the haptic feedback in the steering wheel.

5.1 *Integration of Guardian Angel HMIs*

The Guardian Angel system's effectiveness is assured through a set of well-integrated Human–Machine Interfaces (HMIs) within a real vehicle. This section sheds light on the visual, auditory, and haptic HMIs that collaboratively enhance driver-system interaction, bolstering overall safety.

Visual HMIs:

- An In-vehicle display to the right of the steering wheel provides audio-visual instructions, guiding users through the Guardian Angel's features.
- A Heads-Up Display (HUD) situated behind the steering wheel projects essential road data, ensuring drivers have vital information directly in their line of sight.
- Ambient lighting, facilitated by LEDs in the windshield and steering wheel, creates an immersive environment for relaying cues and alerts.

Auditory HMIs:

- Auditory cues offer proactive alerts about different hazards, heightening driver vigilance and responsiveness.
- The audio tutoring system plays a key role, especially when drivers can't focus on visual displays. It provides dynamic guidance through audio messages, explaining specific actions performed by the Guardian Angel system.

Haptic HMI.

- The steering wheel's haptic feedback employs vibrations to alert drivers about imminent hazards by means of tactile communication.

Further details about each HMI's functions can be found in the public deliverables of the Hadrian project, offering insight into their specific contribution to improve the collaboration between the driver and the Guardian Angel system. Besides, this chapter will only describe in detail the Ambient Lighting HMI and the Haptic HMI (Fig. 4).

5.2 *Ambient Visual HMI*

The visual HMI consists of ambient light indicators that act in addition to the set of standard visual dashboard, center-stack, and other in-cabin indicators commonly present in passenger vehicles.

5.2.1 Concept

In contrast to direct visual information, ambient cues transport information in a more subdued and indirect way. Ambient cues can vary from being individual signals that

Fig. 4 View of the internal HMI of guardian angel system

are outside the viewer's center of direct attention, to fully changing the situation that surrounds an individual's actions and points of interest. Within the Guardian Angel, the former concept is used, i.e., lights are strategically positioned to emit visual cues that guide the driver's gaze towards points of interest or attention within the driver's field of vision but outside the direct center of attention. The aim of the ambient visual HMI is to not serve as the primary visual HMI but instead as a set of *secondary indicators*. The ambient visual HMI is intended to enhance and support the driver's situation awareness in terms of driver and system readiness status as well as add an explanatory layer to the vehicle's interventions and the haptic steering wheel feedback. In particular whenever the system intervenes, the light is emitted from the side the vehicle is steering away from. Thereby, it directs the driver's attention towards the cause for the vehicle's intervention. Likewise, the lights indicate and accompany interventions from either side and assist the driver to intuitively distinguish between assisting (side lights on during and off after steering intervention from respective side) and overriding (solid lights on both sides) modes without requiring active attention. Providing this additional layer of information enables a more informed interaction with the vehicle and diminishes the likelihood of unsafe driver behavior caused by resistance to or confusion by the vehicle interventions.

5.2.2 Implementation

The positions of the ambient visual HMI Are shown in Fig. 5.

The windshield is surrounded by three light indicators: two along the left and right a-pillars (wL and wR, respectively), and one at the bottom of the windshield (wC). The steering wheel also contains three indicators, all of them aligned horizontally at the center of the wheel: One left (sL), one right (sR), and one central indicator (sC).

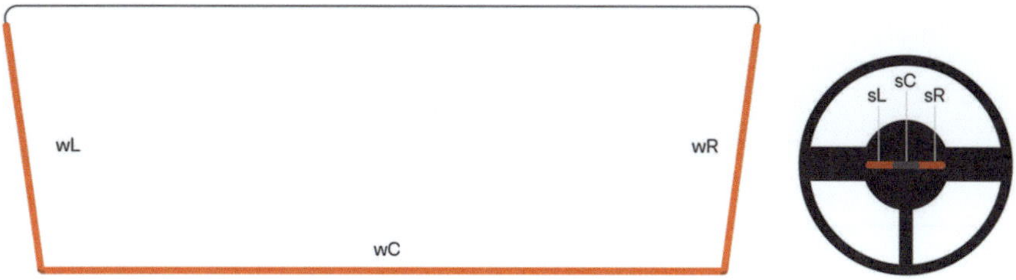

Fig. 5 Overview of the lights in the windshield area (left) and in the steering wheel (right)

Indicators wL and wR are the primary source for indicating vehicle interventions and haptic feedback on the steering wheel, whereas sL and sR serve as the secondary indicators for the same purpose. wC is used to indicate driver distraction and draw attention to the road center. sC is used to indicate the system status.

The indicators are to be realized as light stripes (LED) using simple color-coded light patterns to indicate driver status, system status, and system activity in a binary manner (system acting vs. not acting, driver distracted vs. attentive, etc.). The used colors are green, red, and orange based on the ANSI Z535.1 American National Standard for Safety Colors.[1]

5.2.3 Functionality

Within the Guarding Angel, a discrete set of states for each indicator is defined to effectively communicate driver and system status as well as vehicle interventions (see Fig. 6 for an overview).

The system status indicator at the center of the wheel (sC) is off during regular operation (system active) in order to decrease distraction, as this is the expected state while driving. If the Guardian Angel is switched off, then this is indicated in red (see Fig. 6a). Upon startup, sC flashes in green for a period of 3 s (Fig. 6b) to signal proper activation of the Guardian Angel. It then switches to regular operation mode (sC off).

Indicators wL and sL light up in orange whenever there is a steering intervention from the left (Fig. 6c); the same is true for wR and sR and interventions from the right side (Fig. 6d). This will draw the attention towards the source of the haptic force felt by the driver as well as the likely reason for the intervention (e.g., proximity to oncoming traffic). Thus, it serves as the more appropriate visual supplement as opposed to indicators that would draw the driver's attention *towards* the steering direction instead of *against* it. Indicators wL, wR, sL, and sR light up in unison (Fig. 6e) when the highest level of support is reached, i.e., when the wheel applies the maximum amount of torque or completely overrides the steering.

Driver distraction is indicated only on the central windshield indicator (wC) and also in orange (Fig. 6f). The distraction indicator is active for as long as the driver is

[1] https://www.nema.org/standards/view/american-national-standard-for-safety-colors.

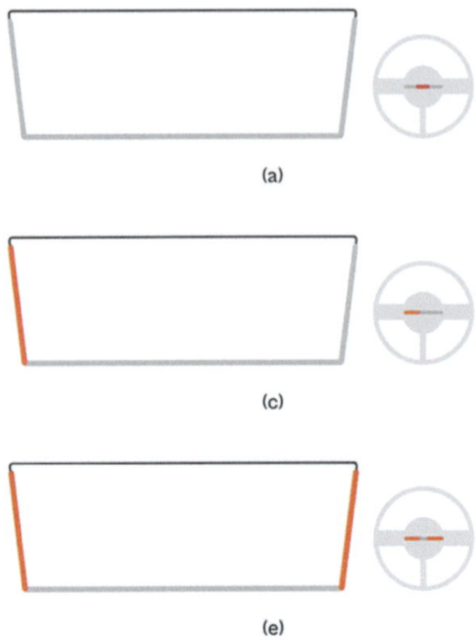

Fig. 6. Overview of light indicator states

detected to be in an inattentive state. For illustrative purposes, the distraction indicator is shown without any other indicators active in Fig. 6 but can be active together with the intervention indicators, depending on whether haptic interventions are caused by or occur in tandem with the driver being distracted.

By using simple color language based on common safety standards, binary state indications, and—apart from the startup system status indication—static (i.e., non-animated) indicators, the ambient visual HMI assists the driver in understanding the Guardian Angel's actions and re-focus their attention in cases of driver distraction. By keeping the communication channels separate and the communication contents to a clear and efficient minimum, the driver is supported to better interact with the vehicle as a whole without running the danger of being itself an additional distractor.

5.3 Haptic Feedback on the Steering Wheel (HAPTIC ICONS)

5.3.1 Concept

Haptic feedback in the steering wheel is a technology that allows drivers to receive information through touch in a fast and efficient way. It can be used to alert the driver of the status of the automated driving system, such as when the system is ready to take over or to transition back to manual control, or to communicate other important information, such as potential hazards on the road (Sarabia et al. 2024). It also provides more detailed information by using different patterns of vibrations, such as the direction of a potential hazard. In vehicles, haptic feedback is a reliable and efficient way to communicate important information to the driver, and it can help to enhance the overall driving experience and improve safety.

5.3.2 Description

Haptic icons are a specific type of haptic feedback, which are brief haptic stimuli that are associated with specific meanings. Similar to language, haptic icons can be considered as "haptic words" that are constructed using haptic phonemes. These phonemes are created using simple waveforms with specific amplitude, frequency, and duration, and they represent attributes of the message such as urgency, attention required, and smoothness.

Haptic icons have several design restrictions and considerations that must be considered (Sarabia et al. 2022). The actuator used has a maximum vibrating frequency of 40 Hz, which is an appropriate value provided the range of perceivable frequencies through touch. Frequencies close to and over 100 Hz are difficult to be perceived by the touch and would probably mask with common driving vibrations, while vibrations below 10 Hz are too slow, and they might be perceived more like an oscillation in the steering wheel. The amplitude must be within a safe range that does not cause rotation or control loss of the steering wheel. Additionally, the waveform used for the phonemes, such as sine or square, must be chosen carefully to ensure that it is easily identifiable and intuitive in use for the driver.

Haptic icons provide a subjective perception in the automation to driver communication. Therefore, subjective studies were performed to obtain some suitable icons.

5.3.3 Functionality

The haptic icons used in this specific context are divided into three different groups: message notifications, take-over requests, and hand-over transitions. In all three cases, even if the waveform varies, they all share the same frequency, 20 Hz.

- Message notifications provide plain information that is not time-critical and is intended for indications with low priority. It is similar to a common phone notification vibration, based on a sinusoidal waveform, and it produces the same stimuli of attention caught (Fig. 7).
- Take-over requests indicate that the automated system can no longer continue driving the vehicle and the driver will need to resume control within a time frame of at most 5 s. It is a repeating notification icon but increasing its amplitude in each iteration. This strength increase causes an urgency feeling in the driver, as if the vehicle was requesting the attention from the driver. The final square pulse defines the end of the haptic icon, so that the driver can interpret that the haptic

feedback ends there, and afterwards, the driving must be resumed by the driver (Fig. 8).

– Hand-over transitions indicate that the automation system is being activated or deactivated. For that, a sinusoidal waveform with increasing amplitude is used. This icon is launched when the driver actively changes the system state pressing a button, so both having a notification icon, or an increasing amplitude could be useful as confirmation. However, the amplitude increase itself provides a feeling of transition, enhancing the intuition of transition. Furthermore, the square pulse provides an initial faster response, to instantly confirm the incoming transition (Fig. 9).

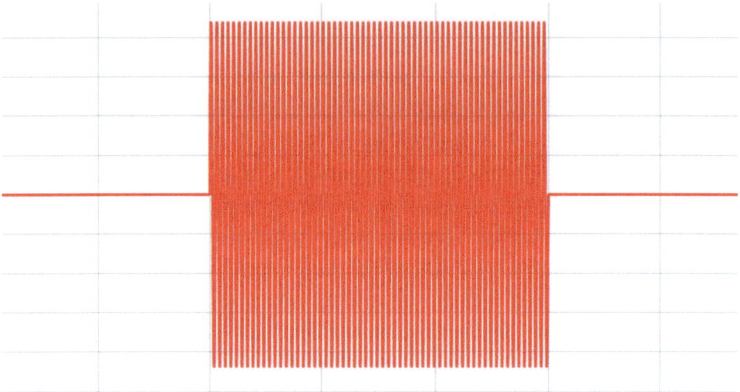

Fig. 7 Waveform of the notification icon

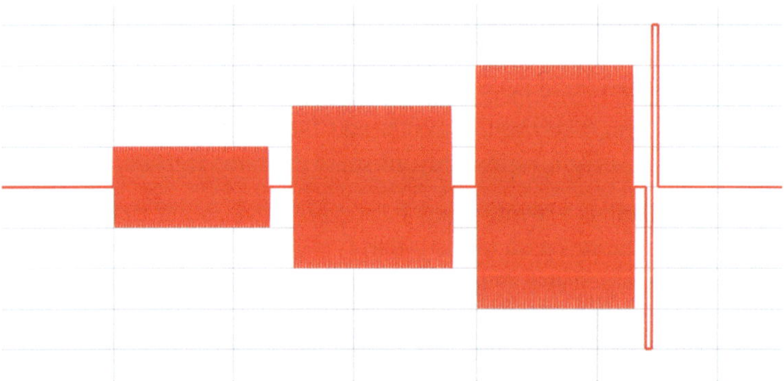

Fig. 8 Waveform of the Takeover request icon

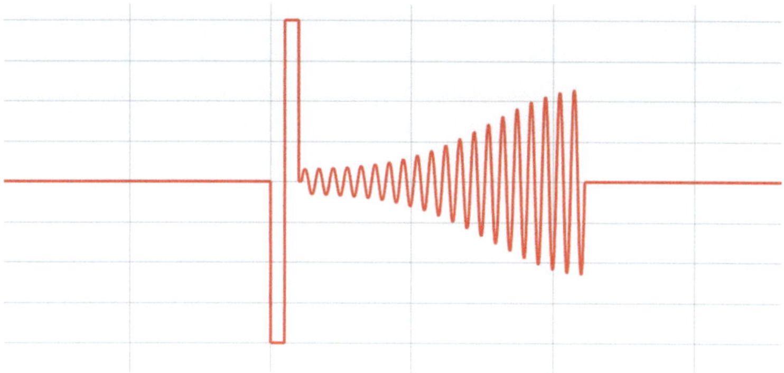

Fig. 9 Waveform of the transition icon

6 Conclusions and Future Works

This chapter introduces the Guardian Angel (GA) concept, a novel human-centered approach to automated driving that prioritizes driver safety. Within the GA system, two mechanisms have been developed and combined to enhance driver safety. The shared controller ensures that the vehicle remains within a predefined safety envelope, while the multimodal Human–Machine Interface (HMI) keeps the driver attentive and aware of the driving situation. This system has undergone development and evaluation through an iterative user-centric process. Multiple user studies have been conducted to assess these functionalities, and the results are detailed in the HADRIAN studies report (Sarabia et al. 2022, 2023, 2024).

Taking the results of the user studies into account, the following conclusions can be drawn:

Regarding the functionality designed to prevent collisions and keep the vehicle on the road, the GA controller has proven effective in increasing safety and enhancing driving performance. Various levels of controller strength were tested to identify the most suitable option, with torque levels up to 6 Nm considered acceptable for lateral corrections and 12 Nm for extreme situations.

When designing haptic icons for steering wheel feedback for the GA, several factors need consideration. Firstly, it can be challenging to differentiate between frequencies, especially in a vibrating vehicle. Therefore, using a constant frequency is the recommended approach, allowing the driver to become accustomed to it. Concerning vibration amplitude, it should be adjusted based on the vehicle's environment. Vibrations should not be too strong, as they can induce panic, nor too soft, as they may go unnoticed amid other vibrations. An adaptive gain, potentially linked to vehicle speed, is suggested to address this issue. Additionally, the duration of haptic icons should be carefully considered. Icons that are too short might be mistaken for road irregularities, while excessively long ones can be uncomfortable and annoying.

To assess these systems, evaluation and validation typically rely on subjective results gathered through questionnaires. Employing standardized questionnaires is essential for ensuring the accuracy and replicability of tests. However, it's important to note that supplementing these questionnaires with methods such as 'speak aloud' methodologies can offer valuable insights that questionnaires might overlook.

As highlighted in the introduction, elderly individuals could significantly benefit from the GA. Improved driving performance facilitated by lateral support and a simplified user interface employing intuitive visual and haptic feedback could notably enhance the confidence of elderly users in ADAS systems. Consequently, forthcoming studies on the GA will concentrate on evaluating these systems specifically for this focus group.

Additionally, the future development of the Guardian Angel (GA) system will also consider new scenarios. As automated driving technologies evolve, human drivers will need to face new situations of collaborative and automated driving. Consequently, advanced ADAS like the GA will continue to be a subject of study and development.

References

Flemish FO, Goodrich KH, Adams AA, Conway SR, Palmer MT, Schutte PC (2003) he H-metaphor as a guideline for vehicle automation and interaction. University of Munich, Munich, Germany. Accessed: May 24, 2021. [Online]. Available: http://www.sti.nasa.gov

Mack M, Stojan R, Bock O, Voelcker-Rehage C (2022) Cognitive-motor multitasking in older adults: a randomized controlled study on the effects of individual differences on training success. BMC Geriatr 22(1):581. https://doi.org/10.1186/s12877-022-03201-5

Makishita H, Matsunaga K (2008) Differences of drivers' reaction times according to age and mental workload. Accid Anal Prev 40(2):567–575. https://doi.org/10.1016/j.aap.2007.08.012

Marcano M, Diaz S, Vaca M, Pérez J, Irigoyen E (2021) Shared control framework and application for European research projects. Adv Intell Syst Comput 1268:657–666. https://doi.org/10.1007/978-3-030-57802-2_63

Marcano M, Diaz S, Perez J, Irigoyen E (2020) A review of shared control for automated vehicles: theory and applications. IEEE Trans Hum-Mach Syst 50(6):475–491. https://doi.org/10.1109/THMS.2020.3017748

Marcano M et al. (2021) From the concept of being 'the boss' to the idea of being 'a team': the adaptive co-pilot as the enabler for a new cooperative framework. Appl Sci 11(15). https://doi.org/10.3390/s24020562

Payyanadan R, Lee J, Grepo L (2018) Challenges for older drivers in urban, suburban, and rural settings. Geriatrics 3(2):14. https://doi.org/10.3390/geriatrics3020014

Sarabia J, Diaz S, Zubizarreta A, Perez J (2022) Design requirements for the definition of haptic messages for automated driving functionalities. In: Proceedings of the 6th international conference on computer-human interaction research and applications, pp 171–178. https://doi.org/10.5220/0011537700003323

Sarabia J, Marcano M, Diaz S, Zubizarreta A, Perez J (2023) Lateral evasive maneuver with shared control algorithm: a simulator study. Sensors 24(2):562. https://doi.org/10.3390/s24020562; https://www.mdpi.com/1424-8220/24/2/562

Sarabia J, Marcano M, Vaca M, Diaz S, Perez J, Zubizarreta A (2024) User-centric iterative development of haptic steering wheel feedback for automated driving applications. Transp Res Part C Emerg Technol

de Winter J, Abbink D, Petermeijer SM (2022) (14) (PDF) Shared control versus traded control in driving: a debate around automation pitfalls. https://www.researchgate.net/publication/357255361_Shared_control_versus_traded_control_in_driving_A_debate_around_automation_pitfalls. Accessed 27 April 2022

Results of Two Demonstrations of Holistic Solutions for Automated Vehicles to Increase Usefulness and Safety

Peter Moertl, Nikolai Ebinger, Cyril Marx, Selim Solmaz, Christoph Pilz, Joseba Sarabia, Sergio Diaz, Mauricio Sandoval, Marios Sekadakis, and Srdan Letina

Abstract In this chapter we describe the conduct and results of two field-demonstration studies where the HADRIAN innovations were integrated into real vehicles and their effectiveness in terms of the human driver role were evaluated. In the first field-demonstration, twelve participants evaluated HADRIAN innovations that were intended to improve the safety and comfort of using automated driving at SAE L2 and L3. The study was performed in a passenger vehicle in an open road environment where participants compared interactions with a baseline automated vehicle with the HADRIAN innovations. In the second field-demonstration, ten participants experienced HADRIAN innovations that were intended to facilitate older drivers in a small passenger vehicle while driving on a test track in Spain. The results of both studies confirmed the key assumptions of the HADRIAN approach and identified limits and opportunities. We discuss these main lessons learned and conclude with what these findings tell us about the benefits and problems of adopting a holistic, user-centered approach for automated driving solutions.

Keywords Human-centered design · Human-systems integration · Automated driving · CCAM · fluid interactions

P. Moertl (✉) · N. Ebinger · C. Marx · S. Solmaz · C. Pilz
Virtual Vehicle Research GmbH, Graz, Austria
e-mail: Peter.Moertl@v2c2.at

J. Sarabia · S. Diaz · M. Sandoval
Tecnalia, Derio, Spain

M. Sekadakis
Northern Technical University of Athens, Athens, Greece

S. Letina
ASFINAG, Vienna, Austria

P. Moertl and B. Brandstaetter (eds.), *Shaping Automated Driving to Achieve Societal Mobility Needs*, Lecture Notes in Mobility,
https://doi.org/10.1007/978-3-031-52550-6_5

1 Introduction

The previous chapters have presented the HADRIAN solutions that were developed in the HADRIAN project to increase the acceptance, comfort, and safety of automated driving (AD) for various types of drivers and users. These innovations had been previously evaluated in driving simulation studies. In this chapter we describe the studies that addressed following two main questions: Can the HADRIAN innovations as described in chapters two and three work effectively in real vehicles? And do these innovations also provide the benefits as we would expect them from the various driving simulation studies? To address these two main questions, two field demonstrations were conducted. Demonstration 1 evaluated the HADRIAN solutions that were described in Chap. 3 of this book and were intended to help drivers use AD at SAE L2 and L3 (SAE International 2021). Demonstration 2 evaluated the guardian angel functionality that was described in Chap. 4.

2 Demonstration 1: Improving Automated Driving Level 2 and 3

The first demonstration measured the feasibility and benefits of the HADRIAN innovations to improve the safety, comfort, and effectiveness of drivers' use of AD at SAE L2 and L3. The motivation for these HADRIAN innovations was described in Chap. 1 and their evaluation in driving simulation studies was described in Chap. 3. The HADRIAN innovations that were evaluated in the field demonstration were based on two extensions of Driving Automation Systems (DAS) compared to conventional DAS: First, road infrastructure information was actively linked to the vehicle to make automated driving more predictable to the driver. Secondly, the driver was made active part of the DAS through fluid interactions that were given depending on the state and behavior of the driver. Thirdly, drivers were tutored about the DAS using an in-vehicle tutoring system that also offered fluid, active recommendations to further improve driver competences during transitions from SAE L3 driving to manual driving. In addition, the vehicle enabled higher levels of automated driving only when the driver successfully demonstrated the necessary skills and competences to use the lower level.

The HADRIAN innovations were integrated in an AD research vehicle, a Ford Mondeo that was equipped with advanced automated driving functions that allowed to operate in SAE L2 and L3, given clearly specified safety regulations that included a safety driver who was at all times responsible for the safety of the vehicle and its occupants. In this demonstration vehicle, two types of Human–Machine Interfaces (HMI) were implemented: one that resembled a baseline DAS without enhanced predictability or driver inclusion. This functionally resembles implementations of SAE L2 and L3 displays in commercially available vehicles. The second type of HMI,

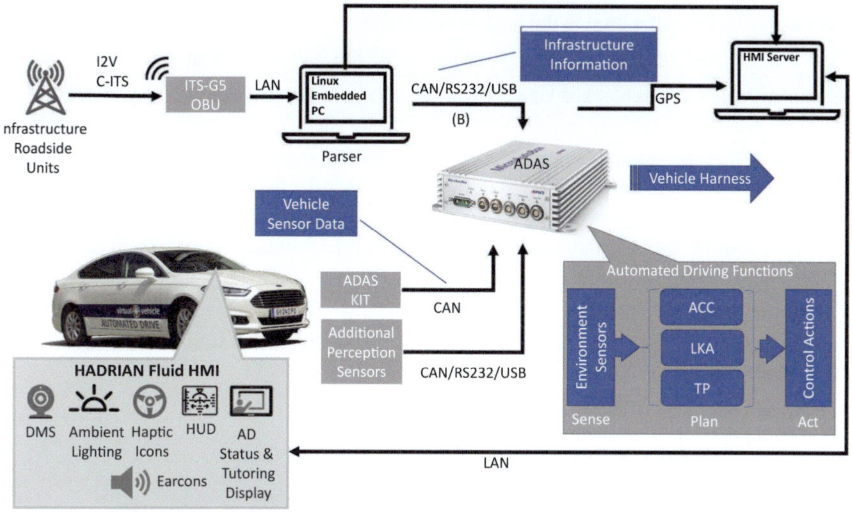

Fig. 1 Demonstration vehicle information architecture

implemented the HADRIAN innovations on five in-vehicle displays and control components that are described in the following subsection.

2.1 *Hadrian Das*

Part of the HADRIAN DAS were a set of infrastructure road side units that sent critical information via C-ITS to the demonstration vehicle, see upper left in Fig. 1. The used C-ITS messages are listed in Table 1. Once received by the vehicle, the messages were parsed and decoded and with information from the vehicle's road environment sensors and in-cabin sensors sent to the ADAS sub-system (in the center of Fig. 1). There they were processed for use by the automated driving functions. Outputs were then sent to the HMI server that managed the display of information on the vehicle displays (lower left hand in Fig. 1) that are described in the next section.

The C-ITS information that was sent from the road side units to the vehicle is listed in Table 1. The used C-ITS message types were all already deployed in the Austrian C-ITS network.

2.2 *HADRIAN HMI in Vehicle 1*

The HADRIAN HMI consisted of five components as shown in Fig. 2.

Table 1 C-ITS messages used in the field-demonstrations

Communicated information	Scenario	C-ITS message format
Terminate ADL 2 due to blocked lane ahead	1	DENM Road Works Warning (4.2.1*)
Continue ADL 3 through upcoming lane closure and lane ahead	2	DENM Road Works Warning (4.2.1) and DENM Hazardous Location Notification (3.2.7)
Information for the driver about ADL 3 portions on planned trip	3	IVIM Automated Vehicle Guidance (5.2)
Unexpected termination of ADL 3	4	DENM Road Works Warning (4.2.1)
Support multiple planned transitions between driving at SAE L2 and L3	5	IVIM Automated Vehicle Guidance (5.2)

*Section number in the C-Roads specification (C-Roads 2021)

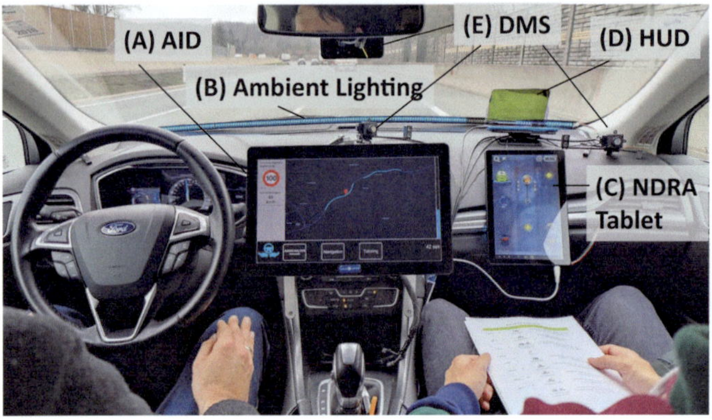

Fig. 2 The HADRIAN HMI was oriented toward the evaluating participant on the passenger seat

The (A) Automation Information Display (AID) in the center console showed all information related to the automated driving on a central location. The AID was divided into four areas that give the participant access to all pertinent information related to automated driving, see Fig. 3. The currently active automated driving level was shown using the symbols of hands and steering wheel, see lower left in Figs. 3 and 4. Thereby, the turquoise hands below the steering wheel indicate that SAE L3 is currently active. The icon for SAE L2 was shown in white, thereby highlighting the difference in the drivers responsibilities. No icon was shown for manual driving as this was found to be potentially confusing by participants in earlier studies. In addition, the remaining duration of the automated drive was displayed in minutes on the lower right of the AID as well as a graphical countdown timer on the right side. The center display of the AID either gave the participants access to the tutoring application or, if desired, showed the navigation display with the current position of the vehicle. On the navigation display, the stretch of the road where automated

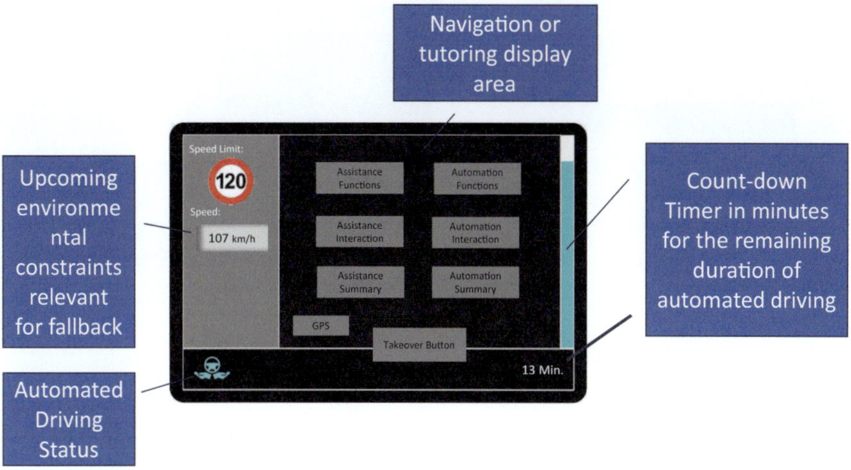

Fig. 3 Display areas of the automation information display (AID)

Fig. 4 Used symbols for automated driving levels

driving was possible was depicted on a map. Finally, the left-hand side of the AID displayed pertinent information about the situation, namely the current speed limit and whether a lane change was coming up.

In the bottom area of the AID were touch-screen buttons that allowed participant to initiate and terminate automated driving. Also, the tutoring application and the navigation display respectively, could be selected. Finally, on the left side of the AID, pertinent contextual information about the environment was displayed: the current speed limit, the vehicle's current speed, as well as any required upcoming maneuvers when transitioning back to manual driving.

Ambient lights ((B) in Fig. 2) indicated the current status and availability of automated driving at SAE L3. The ambient lights also indicated when an unsafe driver state was detected in manual or SAE L2 driving. A driver state was identified as unsafe when the driver glanced away from the road for too long or removed the hands from the steering wheel for too long. In that case the center part of the LED strip flashed in red. Thereby, the permissible amount of glancing away depended on the current level of automation such that in manual mode an alert was shown when the driver glanced away from the road for >2 s or moved both hands from the wheel for more than 1 s. When driving in SAE L2, the alert was shown when the participants glanced away from the road for >3 s or moved both hands from the wheel for more

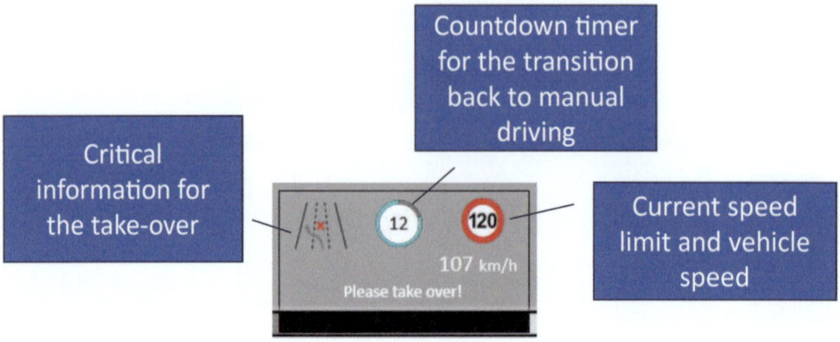

Fig. 5 Areas on the head-up display

than 3 s. No driver state alerts were shown when driving in SAE L3 as this level imposes no monitoring requirements on the driver.

A computer tablet (C) was mounted to allow the participants to engage in non-driving related activities while at the same not having to hand-hold a separate device for this purpose.

A head-up display (HUD), (D in Fig. 2) provided the participant with safety critical information dependent on the currently active automated driving mode. The information that was displayed on the HUD was minimized to avoid cluttering and showed only contextually critical information. Specifically, the HUD showed the currently active AD Level, the current speed limit, as well as the next upcoming maneuver and the remaining time to take back control during the take-over. Thereby, the HADRIAN DAS allowed drivers a maximum of 15 s to take back control from SAE L3 and 5 s from SAE L2 (see Shi and Bengler 2022) (Fig. 5).

The components (A–D) were connected to a driver monitoring system (DMS), (E) in Fig. 2, that observed the driver with three RGB cameras.

2.3 *Information Elements in the Vehicle 1 HADRIAN HMI*

Overall, the HADRIAN HMI display elements served to (1) make the automated drive more predictable to the driver, to (2) enhance their mode awareness of the currently active state of the automated driving system, to (3) give dynamic, driver- and environmentally adjusted feedback how the drivers needed to use the automated driving function and to (4) over time improve their competences and skills to safely handle them. Each of these information elements are described next in more detail.

2.3.1 Predicted Duration of the Automated Drive

Three types of AD predictability were provided: (1) the expected availability of the AD function was shown to drivers prior to their trip so that they could better plan their

NDRA during the trip. (2) During the automated drive, the remaining duration of the automated drive was displayed to help them prepare for a take-over. And (3), after a take-over request (TOR) was displayed to the driver, the amount of time available for that take-over was also displayed.

2.3.2 Increased Mode Awareness Displays

Ambient lights were illuminated in turquoise when automated driving at SAE L3 was currently active or whether a transition back to manual driving was necessary. Also, the computer tablet (C in Fig. 2) that was foreseen for use by the driver for non-driving related activities was illuminated by LEDs. In addition, the AID provided automated driving status of the current active ADL and upcoming changes of the ADL, including all AD predictability information. Finally, a HUD provided time critical information to the driver concerning upcoming maneuvers and ADL transitions.

2.3.3 Dynamic Feedback Depending on Driver Attention Allocation

The DMS observed the driver and triggered warnings and instructions as needed to help the driver achieve safe handling of the vehicle. During manual driving or driving at SAE L2, the driver was warned if his or her gaze was not centered on the road environment for too long. During manual driving, the warning was issued when the driver's gaze was detected off the road environment for more than 3 s, during SAE L2 driving, this period was set to 6 s. During automated driving at SAE L3, no such warnings were given. However, during the completion of the driver's take-over from automated back to manual driving, the DMS observed the driver's attention allocation to three safety critical areas: (1) the road ahead, (2) rear-view mirrors, and (3) the current speed on the speedometer. If during the take-over, the driver had only insufficiently scanned these areas, the DMS instructed the driver via auditory message after the take-over about how to improve the driver's environmental scanning prior to taking-over.

2.3.4 An Interactive Driver Tutoring Application Helps the Driver Understand and Use the Automated Driving System

In today's AD vehicles, drivers learn about the automated driving functions primarily using their printed or increasingly, digital vehicle manuals. Because many drivers do not sufficiently use or understand these manuals (Oviedo-Trespalacios et al. 2021), an interactive driver tutoring application was implemented in HADRIAN to help drivers acquire and dynamically improve the needed competences and knowledge to successfully use the AD vehicle. The tutoring application provided instructional information primarily before the drive and informed the driver about the functioning of the automated driving system as well as his or her responsibilities and how to safely use the automated driving system. In addition, the tutoring application reminded the drivers during the initial usage periods about the most important lessons. More details on the HADRIAN tutoring applied in this study can be found in Ebinger et al. (2023) and Chap. 3 in this book.

2.3.5 Baseline HMI

In contrast to the HADRIAN DAS that included the just described HMI, in the baseline HMI the AID (A in Fig. 2), did not display any predictability or tutoring information. Instead, the baseline HMI only displayed the icons depicting the current mode of automated driving (see Fig. 4) Also, this information was not displayed on the HUD, (D in Fig. 2), there were no ambient lights (B in Fig. 2.), and drivers/ participants did not receive any dynamic feedback to help improve their performance. In the baseline, participants informed themselves about the DAS reading a printed manual that described the AD functions.

2.4 Method

The demonstration drives were conducted between March–April 2023. Twelve participants (6 females and 6 males) with a mean age of 37 years (with a standard deviation, SD = 10.52) participated in the study. They had on average 19 years of driving experience (SD = 9.78) and drove on average about 12 000 km in the previous year. Four participants reported to drive daily, three to drive more than once a week, two to drive more than once a month and three to drive less than once a month. Seven participants had previous experience with cruise control, four with active ACC (between 200 and 20 000 km), and two had pre-experience with active lane-keeping assistant (200 and 500km). None of them had previous experience with a lane change assistant. Participants were trained as evaluators prior to the study to be able to better consider the critical aspects of driving automation systems and also account for the fact that, due to safety reasons, they could not drive the vehicle by themselves. Especially trained safety drivers drove the vehicle and engaged and disengaged the driving functions to ensure the safety of the demonstration. This was necessary to meet the safety regulations of the local and national authorities to conduct the demonstration on public roads. Participants were asked to evaluate the two different types of DAS: a HADRIAN DAS and a baseline DAS.

Prior to the study, the participants received two evaluator training sessions. The two sessions were separated to allow participants to better digest the relative complexity of the evaluation and allow for better understanding. The first training session was held on a day prior to the demonstration drives. Here the participants were briefed on the purposes of the evaluation as well as the differences between the investigated automated driving levels and corresponding human responsibilities. Most importantly, participants received guidelines on how to evaluate the system from their perspective on the passenger seat and were encouraged to give fair and unbiased evaluations. Specifically, the participants were informed that they would experience two different DAS systems, referred as "ADAS A and B", which were in fact, the HADRIAN and the baseline DAS respectively. At the end of this first training sessions, the participants were asked to sign the informed consent form. With their agreement to participate in the study they were then invited to the second day when the demonstration drives were conducted. At this second training session on the day of the demonstration drivers, participants were reminded of the material that they had already received on the first day and their the purpose and evaluation process again clarified.

Then the actual demonstration drives started. For each drive, one of two specially trained safety drivers operated the AD vehicle during manual and automated operations to ensure the safety of the demonstration in the real road environment. The evaluator was seated on the right-front seat and evaluated the system "as if" actually using the system.

2.4.1 Procedure

The evaluation procedure followed the steps shown in Fig. 6. On day 2 the participants entered the vehicle and the vehicle instrumentation was explained. After the familiarization, the safety driver drove the vehicle to the designated parking spot where the experimenter prepared the participants for the first drive. Each drive was on a motorway and about 10 km long, only the last drive was about 20 km long. Participants experienced each drive in a specific scenario that is described in more detail below. The drives were organized in two blocks: the first block consisted of three drives in the baseline condition and the second block of five drives in the HADRIAN condition. The first block was shortened to three drives to keep the duration of the demonstrations manageable for participants and the involved personnel as well as reduce fatigue effects. After each drive, the vehicle was stopped at a safe location and the experimenter asked the participants about their experiences using a semi-structured questionnaire. After the first block of drives in the baseline condition, the participants were asked to complete a survey to collect their experiences with the baseline DAS. In the second block of drives, the participants were introduced to the HADRIAN DAS. Participants watched a tutoring video that explained the SAE L2 automation before starting the next drive. After that drive in SAE L2, participants watched again a tutoring video which explained SAE L3 automation before starting the following drives. Prior to the last drive, participants watched a video explaining the use of both SAE L2 and SAE L3 automation. After completing each block of drives, the participants were asked to complete a final survey and answer a set of questions to evaluate their experiences. The experimenter then asked open questions about the participant's experiences and for any final feedback. Then participants received their compensation of 200 Euros for their time and effort and were released.

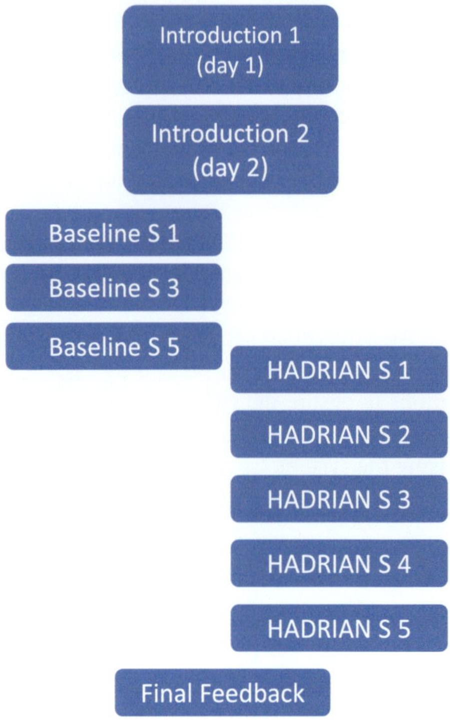

Fig. 6 Sequence of evaluation procedure

2.4.2 Materials

Participants compared the baseline and the HADRIAN DAS on a scale from −10 (strong preference for the baseline version) to +10 (strong preference for HADRIAN version) on a self-developed questionnaire. Additionally, the experimenter asked the participants verbally to explain their rating on following dimensions:

- Safety: "Do you think drivers find it safer to drive with the BASELINE or HADRIAN DAS?"
- Joy: "Did you find it more enjoyable to play on the tablet when using the BASELINE or HADRIAN DAS?"
- Usefulness: "Did you find it more useful in your private life to be able to use the BASELINE or HADRIAN DAS?"
- Assistance: "Did you find the introduction to driving assistance and driving automation more effective in the BASELINE or HADRIAN DAS?"
- Assistance displays: "Did you find the display of driving assistance and driving automation status more effective in the BASELINE or HADRIAN DAS?"
- Take-over: "Did you find the request to take back control more effective in the BASELNE or HADRIAN DAS?"

In addition to the survey, the experimenter asked the participants verbally after each block about their experiences with the tutoring application and about any perceived privacy issues and recorded their responses. After each block the participants also rated the visibility and usefulness of the experienced HMI elements on a scale from 1 (not at all) to 10 (fully).

2.4.3 Demonstration Scenarios

Five scenarios were selected with the intent to trigger some of the operational difficulties that drivers experience when transitioning between automated driving and manual driving. Specifically, the scenarios addressed expected as well as unexpected transitioning from automated driving back to manual driving, continuing automated driving through an unexpected maneuver, and switching between two different automated driving levels. The five scenarios were the following:

- Scenario 1 consisted of a transition back from SAE L2 to manual driving. After driving for a while in SAE L2 on the motorway, the vehicle approached a construction site that required a lane change and termination of SAE L2 and a transition back to manual driving. For this purpose, the HADRIAN DAS received an early message from the road information infrastructure to allow the participant to prepare for the transition back to manual driving; a warning message was shown 5 s prior to the termination of SAE L2. In the baseline DAS, that resembled cars with SAE L2 of today, no such warning was provided.
- Scenario 2 consisted of continuing SAE L3 driving through an unexpected lane change maneuver. The driver has engaged ADL3 while driving on a motorway. As the vehicle approached an upcoming construction area with a closed lane, this information was sent to the vehicle from the road information infrastructure. The vehicle's AD function used this information to plan a lane change maneuver, to execute it, and to display the upcoming lane change on the AID to the participant. The scenario was shown only in the HADRIAN condition.
- Scenario 3 consisted of a planned termination of SAE L3 driving. Before the drive started, the participant viewed the availability of driving at SAE L3 for the upcoming trip on the AID navigation display. This information was sent from the road infrastructure to the HADRIAN vehicle. After starting the drive, the expected duration of driving at SAE L3 remained visible to the participant in the HADRIAN condition and a warning was issued 15 s time prior to the TOR. In the baseline condition no such information was available and participants only received a warning 7 s prior to the TOR. This timing was based on nominal assumptions about the ability of vehicle sensors to pick up an obstacle ahead while traveling at 100 kph.
- Scenario 4 consisted of an unplanned termination of SAE L3 driving. Before the drive started, the evaluator viewed the availability of driving at SAE L3 for the

upcoming trip on the AID navigation display. Again, this information was previously received from the road infrastructure. After starting the drive, this information remained visible on the AID for the evaluator. In this scenario a temporary lane closure required a lane change for which the driver had to switch back to manual driving. This unexpected lane change was placed prior to reaching the expected SAE L3 termination point. Therefore, a message about this unexpected lane change was sent from the local road infrastructure to the HADRIAN vehicle and was displayed to the evaluator to prepare for a timely take-over. Specifically, the participants were given 15 s time to transition back to manual driving.

- Scenario 5 consisted of repeated switching between SAE L2 and L3 during the drive.
- In the HADRIAN condition, the evaluator viewed the availability of driving at SAE L2 and L3 on the AID navigation display before starting the drive. This information was sent from the road infrastructure to the HADRIAN vehicle and remained visible on the AID throughout the drive. No such information was available in the baseline condition. When the vehicle approached the first AD termination point in the HADRIAN condition, the driver received a TOR giving the driver either 5 s in SAE L2 or 15 s in SAE L3 to take back control or transition back to manual driving. When the vehicle approached the first AD termination point in the baseline condition, the driver received no warning prior to the termination of SAE L2 or 7 s to the termination of SAE L3 time to take back control or transition back to manual driving. Following transitions between Automated Driving Levels occurred in this scenario: from manual driving (MD) $\rightarrow$ SAE L 2 $\rightarrow$ MD $\rightarrow$ SAE L3 $\rightarrow$ MD $\rightarrow$ SAE L3 $\rightarrow$ MD $\rightarrow$ SAE L2.

2.5 Results

Overall, the twelve participants compared the baseline system and the HADRIAN DAS on a scale from -10 (preference for the baseline DAS) to $+10$ (preference for the HADRIAN DAS) on six dimensions, the results are shown in Fig. 7. Whiskers indicate standard deviations. Each dimension was tested for significance using t-tests against a 0 rating which would indicate no difference between the two conditions. The t-tests were Bonferroni corrected due to multiple comparisons (Abdi 2007). The tests are statistically significant at the $p = 0.05$ level on all dimensions which indicates a strong preference for the HADRIAN DAS compared to the baseline DAS.

On average, the HADRIAN DAS was rated significantly better than the baseline on all dimensions (see Fig. 7). The participants' verbal responses were then grouped into distinct categories and their frequencies were counted. The most common responses for each of the six dimensions are presented in the following paragraphs, the subsequent numbers in parentheses represent the number of participants with that response.

Participants felt on average safer using the HADRIAN DAS compared to the baseline DAS. Especially, they felt safer with the tutoring system (7). For example,

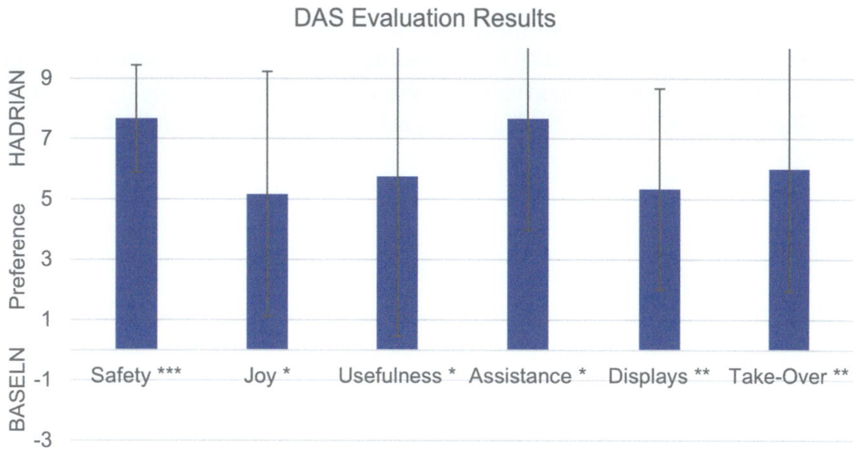

Fig. 7 Comparative ratings between baseline and HADRIAN system (asterisk indicates statistical *p < 0.05, **p < 0.01, ***p < 0.001)

one participant stated, "You feel guided, not too rushed, you have a guide and lead on what to do and when to do it". Participants also referred to several aspects of the tutoring including the pre-drive explanation video (2), situation-sensitive explanations while driving and (5) adaptive feedback after taking over manual control (3). The second most frequent response was that the HADRIAN DAS indicated how long the automation will be available (6). Participants generally felt safer in the HADRIAN condition because the status information was provided in a variety of ways, such as using ambient LED lights (6).

Participants enjoyed playing on the NDRA tablet in the HADRIAN condition more because they knew how much time they had left and they were better prepared to drive manually again (7). Participants also reported that they enjoyed the NDRA more because they felt safer (4). Some drivers did not experience a difference in enjoyment and stated that this was because the automation was reliable and thus allowed them to focus on the NDRA in both DAS anyway (3).

The HADRIAN DAS was rated as more useful than the baseline DAS. Participants explained that this was primarily due to the predictable information about the availability of driving automation (5) and the comprehensive support provided by the HADRIAN tutoring (4).

2.5.1 HMI Component Evaluations

Participants rated the AID in the HADRIAN DAS on average as more effective than the baseline display. Thereby, seven participants mentioned ambient LED lighting as highly effective. In addition, they found the predicted duration of the automated drive as useful for upcoming automation changes (4). Accordingly, this was also reflected in the fact that participants preferred the process of getting back to manual control in HADRIAN, because the information about the duration of the automated drive allowed them to better prepare (8). In addition, the initial tutoring introductions about the take-over process were found to be valuable for actually performing the take-over (4) and participants found the flashing ambient LED lights useful to indicate the TOR (3).

Tutoring

The HADRIAN tutoring was reported as the most important reason for the increased feeling of safety in the HADRIAN condition. Six participants explained that they preferred watching the video presentation of the DAS over the need for reading a text in the baseline. Six participants empathized that they liked that the tutoring repeated important aspects. At the same time, three participants added that the tutoring provided short and understandable explanations. In contrast, one participant did not like the explanations provided by the tutoring while driving. Individual participants also proposed adding additional features such as a quiz and an indication of the remaining video length.

Participants preferred the HADRIAN tutoring instruction compared to the written manual in the baseline DAS primarily because of the initial introduction videos (11). In addition, three participants highlighted that they liked that the tutoring videos repeated the most important information. However, one participant preferred the written information in the baseline over the video in the HADRIAN condition. As indicated above, participants were introduced to the DAS functionality in three incremental steps. First, they experienced SAE L2 automation, then SAE L3 and only in the end did they experience both SAE L2 and L3 automation within the same drive. All participants were positive about this stepwise approach to be introduced to automation, but three participants wanted to limit this restriction to the first time they use the vehicle. When asked if it would be acceptable in their private car, all participants said yes. One participant limited acceptance to the ability to skip the stepwise introduction procedure.

Evaluations of Other HMI Elements

While the focus of this study was to identify the feasibility and benefits of the main HADRIAN features in a real-world environment, participants also evaluated the specific HMI elements on a scale from 1 to 10 to identify possibilities for improvement. All participants reported that the audio signals were clearly audible, and seven participants rated them as distinguishable. Ten participants rated the icons to be distinguishable and eleven participants rated them to be understandable. Nine participants found the ambient lighting to be visible and eight found it useful. Eleven participants rated the messages on the AID display as visible and useful. Six participants rated the HUD as visible and eight rated it as useful.

In summary, the results point to the benefits of tutoring and predictability information as key technological improvements for automated driving on all of the six measured acceptance dimensions. The actual HMI that was used to implement these HADRIAN innovations appeared to be adequate though participants would have preferred the HUD to be more visible.

Finally, participants did not see privacy as a particularly important issue, resulting in a relatively low average importance rating of 2.55 (SD = 2.39) on a scale from 1 (not at all problematic) to 10 (very problematic). Additionally, when asked about possibilities to reduce the impact on privacy that could be taken to make it less of a problem, participants mentioned limiting the storage and sharing of data. Only one evaluator saw a general problem with in-car video recordings but did not provide further information on that.

2.6 Conclusions

Demonstration 1 evaluated the technical feasibility of implementing a holistic DAS that includes the driver as well as the road information infrastructure. The benefits of such holistic DAS were associated statistically significantly with more perceived user benefits than a baseline vehicle. Thereby, the holistic DAS was able to increase the predictability of automated driving so that drivers could plan their activities during the AD trip and better prepare to safely take back manual driving control. Also adaptive, fluid interactions between drivers and vehicles were positively evaluated: the vehicle adapted to the needs of the driver and issued recommendations only when needed. And thirdly, the holistic DAS allowed drivers to increase their AD competences without having to read lengthy printed manuals. Also, the demonstrations showed specific methods to improve automated driving mode awareness with a head-up display and ambient lighting. A tutoring application further allowed drivers to build their competences more comfortably and more effectively to use the AD vehicle safely.

Technical Conclusions

The demonstrations showed that infrastructure-assisted automated driving functions with standard C-ITS messages are realizable to not only improve the safety and comfort of DAS but also to increase their acceptance and their perceived benefit. Such infrastructure-to-vehicle communication could extend the predictability of AD vehicles by overcoming limitations of onboard sensors to detect unexpected changes of the ODD. However, the preparations for the demonstrations have also shown that generating appropriate C-ITS messages is still largely a manual process; specific C-ITS message must be created by a communication engineer for specific events such as lane closures or obstructions. The manual development process not only includes the selection of the appropriate message type but also the geographic information of the event itself, including the regions of applicability and visibility for a specific vehicle to determine whether a message is applicable. The necessity of such manual process is currently a major limitation for wide-spread use of road-infrastructure information. To move toward the promises of C-ITS to improve AD vehicle operations, an automatic or semi-automatic pipeline for generating such infrastructure assistance messages will be required.

In addition, during the demonstration preparations, we identified limitations to implement and use the openly specified C-ITS messages with non-proprietary tools. The only currently free and open-source tool[1] to parse and decode C-ITS messages cannot handle minor specifics of C-ITS messages, such as the standard IVIMv2 message. While a standard for V2X communication exists, no ready-to-use library in C++ or Python currently exists and therefore has to be created for specific applications. This also limits the use of C-ITS messages by a wider community and restricts applications to research rather than wide-spread adoption. To address this, we think that an open encoder library should be adopted to enable effective exchanges for a wider development community. Such open libraries are common for other types of data exchange, such as the MQTT[2] middleware for Internet of Things (IOT) applications. This MQQT middleware is available for a vast majority of programming languages. Establishing something similar for V2X communication is a clear direction that should be explored in the future.

Another observation is the need to appropriately prioritize the information that comes from the infrastructure to vehicle communication. At the highest priority are "recommendations" on how the vehicle should maneuver. The execution of such recommendations depends on the given traffic environment. However, and this is an important limitation, todays C-ITS systems include static ground truth such as lane markings, traffic signs, or safety barriers but no dynamic ground truth about the moving traffic elements. Therefore, prior to implementing a maneuvering recommendation, a vehicle has to check the surrounding traffic and plan a safe maneuver for execution of the recommendation. This will require the joint prioritization of vehicle and infrastructure sensor signals which will need to be developed to safely handle specific road traffic situations.

Improving Human Factors of Automated Driving

The methods that were investigated in the HADRIAN project to improve the human factors of transitioning between automated driving levels are effective and technically feasible. Specifically, we were able to show that by making AD more predictable and transparent to the driver, we can unlock the potential of AD vehicles to not only become safer but also enable a main benefit mechanism for SAE L3 vehicles: drivers could plan their NDRA's on a trip and in this way experience them more positively and effectively. Also, we showed the benefits of directly including the human driver as part of the system. In this way, the driver becomes a reliable partner to the driving automation system who is helped with clarifications and warnings if needed to lead them toward safe interactions with the AD vehicle.

[1] https://github.com/brchiu/asn1c/tree/velichkov_s1ap_plus_option_group_plus_adding_trailing_ull.

[2] https://mqtt.org/.

It is important that these innovations are not purely on the human-vehicle interaction level but go deeper into the functionality of the AD vehicle with legal and social implications. The HADRIAN approach specifies driver responsibilities as well as road information infrastructure and vehicle responsibilities toward achieving a high level of predictability and performance. In this way overall performance levels could be guaranteed, an important precondition for wide-spread adoption of automated driving. Therefore, we can confirm that improving the usability and acceptability of automated driving does not solely rely on the definition of the interface between humans and vehicles but requires to include road infrastructure information as well.

3 Demonstration 2: Guardian Angel

Demonstration 2 addressed a different HADRIAN innovation that was introduced in Chap. 4. The Guardian Angel concept represents a safety system focused on improving manual driving. This system has been designed to offer support to drivers with diminished driving skills, with a particular emphasis on elderly drivers who may have experienced a reduction in their driving privileges. The Guardian Angel system operates in conjunction with the driver's manual control, enhancing safety measures while preserving the individual's sense of autonomy and self-assurance.

The responsibilities of the driver remain unaltered when driving manually with the Guardian Angel system. However, the system discretely intervenes in situa tions deemed unsafe, adhering to the principle of minimal intervention. Importantly, full control is retained by the driver at all times, ensuring broad user acceptance, particularly among elderly drivers who value their independence.

The Guardian Angel system continually monitors both the vehicle's surroundings and its interior to detect potential hazards and respond appropriately. System interventions are triggered incrementally, taking into consideration external risks, such as imminent collisions or lane departures, as well as internal factors, such as driver distraction.

This chapter presents the outcomes of an extensive field-study encompassing two distinct scenarios. The first scenario explored distracted driving, where hazards emanated from within the vehicle itself. In this context, the driver's suitability to operate the vehicle was assessed by the system, which evaluated the driver's gaze and hand positioning on the steering wheel.

In the second scenario, the focus shifted to an external hazard, specifically, a situation in which the vehicle encountered a construction site where the drivable road narrowed significantly. This scenario required higher level of driver attentiveness as negotiating and maintaining the vehicle within the narrower track posed a more challenging task. Depending on these factors and the vehicle's road position, precise lateral corrections were applied by the Guardian Angel's steering system to ensure safety.

3.1 Demonstration Vehicle Description

The vehicle used for these tests was a Renault Twizy, a compact electric vehicle for single-passenger transportation. It has been specially equipped to function as an automated vehicle, integrating a range of advanced technologies to ensure safe and precise control. The localization and path following of the vehicle are made possible through the sensors, computing equipment and actuators.

In terms of localization, the vehicle utilizes a DGPS[3] (Differential Global Positioning System) for highly accurate positioning. To further enhance precision and speed, it integrates its position data with an inertial unit, achieving an update rate of 100 Hz. This combination of technologies ensures rapid and precise location tracking.

When it comes to path following, the system incorporates a repository of pre-planned routes and trajectories based on UTM (Universal Transverse Mercator) coordinates, which are stored locally. These trajectories serve as a reference for guiding the vehicle along its intended route with high reliability.

The core of the automation system is an industrial embedded PC, responsible for receiving data from the localization devices in real-time. This PC processes the incoming data and carries out the decision-making and control functions for the vehicle. These functions encompass generating trajectories and ensuring precise path following.

To facilitate communication between the PC and the vehicle's actuators, a PLC[4] (Programmable Logic Controller) operating on a CAN (Controller Area Network) network has been implemented. This network enables seamless interaction, allowing the PC to interface with specific actuators responsible for controlling the steering wheel and pedals, thereby enabling precise control over the vehicle's movements.

With such systems, the vehicle can perform various aspects of the dynamic driving task. This includes the longitudinal and lateral control of the vehicle, planning routes, detecting road associated risks and avoiding them, maintaining situational awareness and executing emergency maneuvers like emergency stops.

3.2 Description of In-Vehicle Innovations

Apart from the lateral controller designed to support the driver, which is extensively described in Chap. 4, the in-vehicle advancements can be categorized into two distinct groups: innovative HMIs (Human–Machine Interfaces) designed to convey information to the driver, and the DMS (Driver Monitoring System), which provides the automated system with real time data about the driver. This dual functionality enables

---

[3] https://www.oxts.com/rt2004-another-cost-effective-precision-solution/.

[4] https://www.ifm.com/es/es/product/CR2530?_gl=1*1khi1pa*_up*MQ..&gclid=Cj0KCQjwj5mpBhDJARIsAOVjBdqjaInH9yrGUg0qT83Emtdn34LcoDZOEnss16i8boOahR565eRVBdIaAhaJEALw_wcB.

a two-way communication channel between the two driving agents—the driver and the automated system—making the Guardian Angel a collaborative system.

3.3 *The Guardian Angel HMI*

The guardian Angel HMI is composed of various individual components covering different communication channels. Those are showed in Fig. 8 from the driver perspective.

Visual HMIs consist of the ambient LEDs installed in the windshield and in the steering wheel, the Heads-Up Display, and the lateral in-vehicle display, where the pre-drive tutoring is displayed.

Auditory HMIs consist of the sound warning system and the during-drive tutoring system. The latter provides an audio message explaining the previous correction that the vehicle made to the driver. The pre and during-drive tutoring systems support the driving assistance similarly to what it is described in Sect. 2.3.4.

The haptic HMI consists of haptic feedback provided through the steering wheel to the hands of the driver.

3.3.1 Ambient Lighting HMI

The ambient light indicators within the Guardian Angel system serve multiple key functions that enhance driving safety and driver awareness. Primarily, these indicators inform the driver of the current driving mode and accompany mode transitions, serving as a permanent reference point throughout the driving experience.

Fig. 8 Guardian angel's innovative HMIs: an overview from the driver perspective

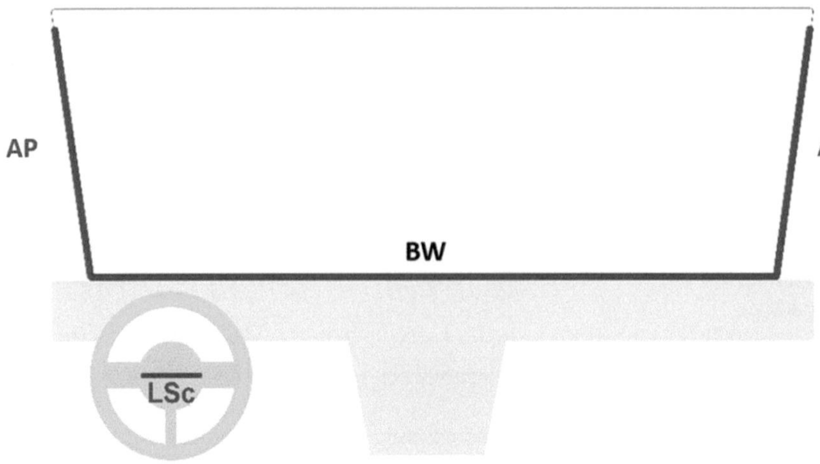

Fig. 9 Ambient lighting viewing areas. AP: a-pillar light indicator; BW: Front indicator Below the Windshield; LSc: Light Steering wheel indicator in the center

In the context of the Guardian Angel application, the primary objective of these ambient light indicators is not merely to indicate mode transitions but to act as warnings for potential driving hazards. To effectively capture the driver's attention in critical situations, the indicators employ an eye-catching orange/yellow color scheme. These indicators are strategically positioned within the vehicle, as shown in Fig. 9:

1. Front Indicator Below the Windshield: This indicator is situated at the front of the vehicle, just below the windshield. As shown in Fig. 8, it is the same LED strip for this indicator and those in the A-Pillars, however, they are controlled independently. Its purpose is to alert the driver to potential hazards related to road departure or driver distraction.
2. A-Pillar Indicators: Two indicators are mounted on each side along the A-pillars. These indicators play a dual role:
   a. They convey the level of authority executed by the haptic steering wheel, offering a visual supplement to the haptic feedback.
   b. They signal external hazards or reasons for the activation of the Guardian Angel, directing the driver's attention toward them.
3. Steering Wheel LEDs: The steering wheel is equipped with LEDs that provide crucial information related to the Guardian Angel's state of activation. The central strip on the steering wheel serves as an indicator of driver attentiveness, offering visible feedback in case of driver distraction. In the event of prolonged distraction, the middle part of the steering wheel LEDs replicates information from the LEDs below the windshield. Additionally, the side strips on the steering wheel mirror the A-pillar LED-strips.

Fig. 10 Heads-up display

3.3.2 Heads Up Display HMI

The HUD shown in Fig. 10 supplies data such as the vehicle's speed, the speed limit, and any upcoming construction zones. Additionally, it offers insights into the status of the Guardian Angel, indicating whether it's currently active or inactive. Also, it provides lane change indications if needed, like in the lane narrowing scenario.

3.3.3 Auditory Alarm HMI

In tandem with visual and haptic HMIs, the Guardian Angel system employs auditory cues to support driver awareness. These sounds serve as critical cues, alerting drivers to varying levels of potential hazards and offering alerts with different levels of urgency based on the driver's state or response.

The auditory alerts are triggered during the activation and deactivation of the Guardian Angel system, as well as when the driver removes their hands from the steering wheel. Each of these warnings has a different sound. They are listed in Table 2.

To ensure their effectiveness, a wireless speaker has been strategically installed within the vehicle. Positioned at the front of the driver inside the dashboard, this

Table 2 Auditory signals list

Sounds		Description
1	On	To inform the driver that GA mode is activated
2	Unavailable	To inform the driver that GA mode is deactivated
3	Hands off the steering wheel	To indicate the driver to hold the steering wheel

placement ensures that the driver's attention is immediately drawn to the auditory alerts, despite the absence of an integrated audio system within the vehicle.

3.3.4 Pre-Drive Tutoring HMI

The in-vehicle display hosts the tutoring application that consists of a 4-min video that provides an informative overview of the Guardian Angel's functionalities and supporting systems.

As shown in Fig. 11, the tutoring video explains participants the Guardian Angel concept, including the responsibilities, functionality, and necessary interactions.

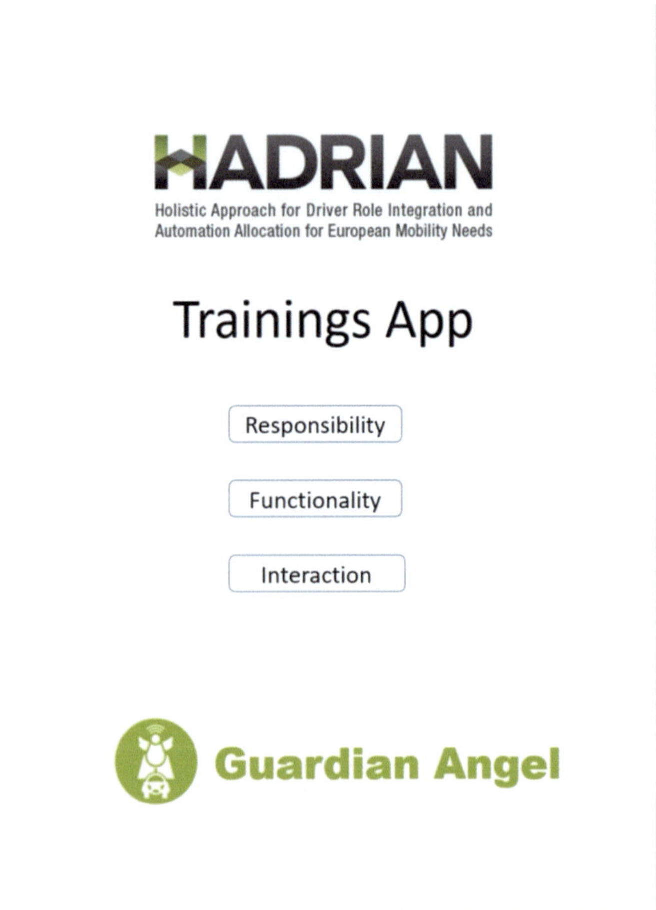

Fig. 11 Introductory frame of the tutoring video

3.3.5 During-Drive Tutoring HMI

The audio voice messages presented during the ride are categorized based on the Guardian Angel system's state. These messages are primarily intended for initial use of the Guardian Angel system to help drivers establish clear associations between visual cues (lights), auditory signals (sounds), and tactile feedback (vibrations) with specific messages. These associations enhance driver understanding and familiarity with the system's alerts.

1. **GA Activated**:
   - **Driver role**: "Guardian Angel activated. Remember, you remain responsible for the drive." This message indicates that the Guardian Angel system is active and reminds the driver that they are still responsible for the vehicle's operation.
2. **GA Active**:
   - **Lateral correction**: "You have approached a lane border. Please, remain within the lane." This message is triggered when the vehicle approaches a lane border, reminding the driver to stay within their lane.
   - **Hands off the steering wheel**: "Please, remember to keep your hands on the steering wheel." This message reminds the driver to keep their hands on the steering wheel when they have removed them.
3. **GA Active and distracted driver**:
   - **Distracted driver**: "You were distracted. Please, remain attentive to the road." In the event of driver distraction, this message prompts the driver to refocus their attention on the road.
   - **Distracted driver 4th time**: "You were distracted. Please, remain attentive to the road." If the driver becomes excessively distracted, this message warns of system misuse and deactivates the Guardian Angel for safety.
   - **Escalation:** "System abuse: excessive distraction time. Vehicle stopped." This message is issued as an escalation in response to excessive distraction, indicating system abuse and leading to a vehicle stop.
   - **Hands off the steering wheel**: "Please, remember to keep your hands on the steering wheel." This message serves as a reminder to the driver to keep their hands on the steering wheel.
4. **GA Active in the lane narrowing**:
   - **Getting into the narrow lane**: "In narrow lanes, the system will provide stronger assistance to center the vehicle."

     When the vehicle enters a narrow lane, this message informs the driver that the system will offer enhanced assistance to maintain proper lane centering.
5. **GA Active in the lane narrowing and with the distracted driver**:
   - **Distracted driver**: "The Guardian Angel is not designed to drive for you. Please, remain attentive to the road."

In narrow lanes and in the presence of driver distraction, this message reminds the driver that the Guardian Angel does not replace their responsibility to drive attentively.

- **Distracted driver 4th time**: "System misuse: too many distractions. Please, stay attentive." Repeated distractions prompt this message, indicating system misuse and urging the driver to remain attentive.
- **Escalation**: "System abuse: excessive distraction time. Vehicle stopped." As an escalation, this message indicates excessive distraction, resulting in system abuse and a vehicle stop.
- **Hands off the steering wheel**: "Please, remember to keep your hands on the steering wheel." This message reiterates the importance of keeping hands on the steering wheel to the driver.

3.3.6 Haptic Feedback HMI

The haptic feedback on the steering wheel introduces two innovations:

Firstly, haptic feedback is employed as a means of communication between the vehicle and the driver. This feedback is transmitted through the driver's hands in the form of waveform patterns, resembling the vibrations experienced with a mobile phone. Within the Guardian Angel's functionality, haptic feedback serves multiple purposes, including notifying the driver of the system's activation and deactivation, as well as encouraging the driver to maintain attentiveness during the drive. When the driver has its hands of the steering wheel, auditory cues are used to let the driver know that gaze distractions were detected.

On the other hand, in hazardous situations, the Guardian Angel assumes control of the steering wheel to ensure the vehicle remains centered within its lane and prevent unintentional lane departures. In these instances, the force exerted on the steering wheel by the Guardian Angel becomes perceptible to the driver's hands, thus qualifying as a form of haptic feedback. This tactile feedback enables the driver to gauge whether the vehicle is actively correcting its trajectory. Such situations may arise due to lane departure scenarios or instances of driver distraction. In response, the controller dynamically adjusts the level of assistance provided and intervenes to rectify the vehicle's path as needed.

3.4 The Guardian Angel Driver Monitoring System

The Driver Monitoring System (DMS) assesses the driver's condition and behavior. It gathers data related to the driver's gaze and hands positions. Gaze recognition is achieved through camera-based technology, while hand position is determined by a combination of camera recognition and a grip sensor integrated into the steering wheel, see Fig. 12. Notably, the camera aimed at the hands also monitors the lateral

Fig. 12 Driver monitoring system (DMS) components: cameras and grip sensor integrated in the steering wheel

display, allowing it to detect any secondary tasks the driver may perform with their right hand on that display.

Within the DMS system, a custom application calculates the driver's fitness level for driving and transmits this information to the main PC. This data informs the Guardian Angel (GA) system, enabling it to customize its support based on the driver's current fitness level.

The necessary equipment for collecting the essential data for the DMS includes:

- A camera positioned to capture the driver's face.
- A camera focused on the steering wheel (hands) and the tablet.
- A steering wheel equipped with a grip sensor.
- A low-level acquisition and communication module (Arduino) integrated within the steering wheel.

3.5 Research Questions

The research questions were chosen to address specific objectives in the context of testing the Guardian Angel system in real-world scenarios.

Research question 1 (RQ1): "Can the Guardian Angel system effectively provide adaptive assistance when a driver is distracted and faced with lane departure risk?"

This research question was chosen to assess the system's ability to respond to one of the most common and dangerous situations on the road: driver distraction. By investigating the system's performance under distracted driving conditions, it helps evaluate the Guardian Angel's potential to enhance safety in real-world scenarios.

Research question 2 (RQ2): "Does the Guardian Angel system demonstrate adaptability in narrow lane situations, considering factors like lane width and driver state?".

Narrow lanes can be challenging to navigate, especially for elderly people, and different drivers may react differently to such conditions. This research question was selected to understand how well the Guardian Angel system can adapt to varying road conditions and driver states, further demonstrating its versatility in enhancing road safety.

Research question 3 (RQ3): "To what extent does the implementation of a multimodal fluid-HMI, including adaptive automation and additional features, enhance the driver's performance and experience during challenging maneuvers?"

This research question focuses on the effectiveness of the multimodal fluid-HMI, an innovative aspect of the Guardian Angel system. It explores how this advanced interface, with features like tutoring, ambient light displays, auditory alarms, and more, impacts the driver's experience and performance. Understanding this is crucial to assess the system's overall usability and the extent to which it can enhance driver capabilities in challenging scenarios.

These research questions collectively provide a comprehensive evaluation of the Guardian Angel system's performance, its adaptability, and its potential to improve road safety in critical driving conditions. They are chosen to address the specific objectives and provide insights into the system's real-world effectiveness.

3.6 Method

Participants experienced the HADRIAN innovations in a real vehicle on a closed test track, where they encountered a series of critical conditions designed to answer the research questions. These conditions included driver distraction (RQ1), negotiating a narrowing lane (RQ2), and experiencing various modes of Guardian Angel intervention with the multimodal HMI (RQ3). The participants underwent these trials with the full Guardian Angel function enabled and also in a baseline condition where none of the innovations were active. For safety, a supervisor was present with an emergency-stopping remote controller, ready to halt the test in case of an emergency. The study intended to address these research questions to evaluate the Guardian Angel system's performance in challenging driving scenarios.

3.6.1 Demonstration Scenarios

The study was composed of two scenarios, focused on the situations addressed by research questions 1 and 2, where the driver faces distraction and lane departure risk from one side, and a narrow lane handling from the other. The scenarios are described in the following subsections.

3.6.2 Distracted Driver Scenario

The Guardian Angel system intervened when the driver was detected to be distracted (as when operating the infotainment system), providing adaptive assistance according to the driver's state and the lane departure risk. During this scenario, the width of the lane was constant for all the route, and it was wide enough to move freely inside the lane. When the participant got close to the lane boundary, the system corrected its trajectory keeping it in the lane. When the driver got distracted, the lateral controller provided a stronger assistance to keep the vehicle centered in the lane. At the same time, the HMI provided the necessary warnings to recover the driving awareness. The driver distraction was induced by requesting the driver to perform a secondary task in the infotainment display.

3.6.3 Lane Narrowing Scenario

The Guardian Angel system intervened when the vehicle was going through a narrow lane, providing an adaptive assistance dependent upon the lane width and the driver state. The narrow lane simulated the case where a construction site blocks part of the road, and the driver has to modify its trajectory and adjust to the space left on the road. When approaching the entrance of the narrow lane, the automated system shifted the trajectory of the vehicle to make the entrance smoother. Regarding the driver, the automated system warned him or her about the incoming narrow lane. During the time the vehicle was crossing the construction zone, the system kept a better control of the vehicle with a consistent lane centering system, as the lane borders were closer. The HMI system kept the driver aware of this additional support until the vehicle departed this zone.

The entrance to the lane narrowing was tested in two different cases, with the driver attentive and distracted. In this scenario, the distraction was induced like in the previous scenario, with a secondary task in the infotainment display. If distracted, the scenario brought 2 different hazards: the external hazard of the more demanding road conditions, and the internal hazard of the driver being distracted. For that reason, in that situation the HMI acted more persuasively to bring back the attention of the driver.

3.6.4 Description of the Test Track

Due to the safety concerns regarding the tests, they took place in a private closed circuit. For each scenario a different loop was set. The test track was 100 m long and 20 m wide. As shown in Fig. 13, the red dotted line shows the circuit for the

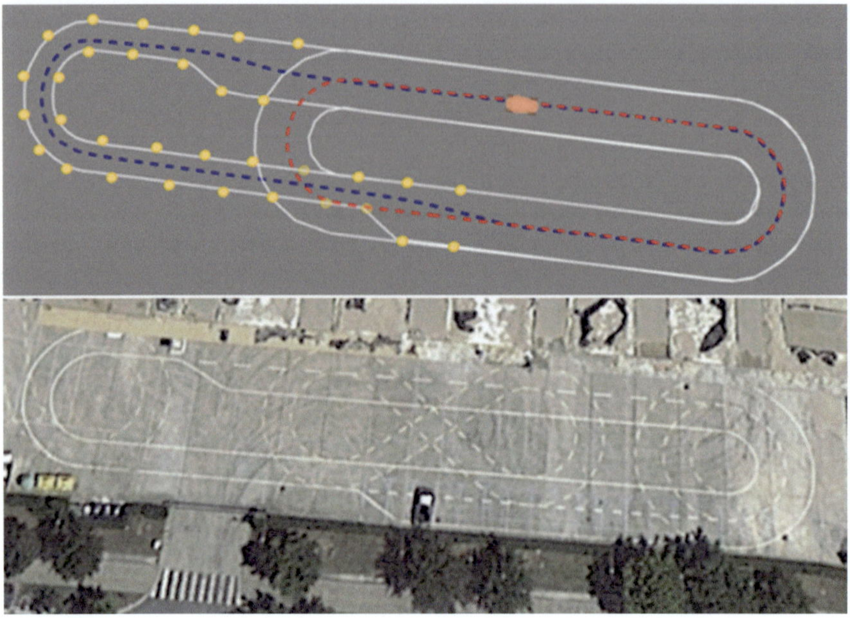

Fig. 13 (Up) Trajectory of the vehicle for each of the scenarios: Red for the distracted driver scenario and blue for the lane narrowing scenario. (Down) Top picture of the actual test track

distracted driver scenario, and the blue dotted line shows the circuit for the lane narrowing scenario. If the tests were performed all in the same direction (clockwise or anticlockwise), the test would be biased, as the drivers would only turn in one direction (left or right). To avoid that, each scenario was performed in one direction, the distracted driver scenario went clockwise and the lane narrowing scenario anticlockwise.

3.6.5 Test Procedure and Instructions

In this section, the test procedure and instructions for the demonstrator tests of the Guardian Angel system are outlined. Participants received an introduction to the vehicle and its safety measures before engaging in familiarization activities. Two distinct scenarios were orchestrated to assess specific system functionalities. *Test Scenario 1* focused on evaluating the system's responses to distracted driving scenarios, while *Test Scenario 2* centered around lane narrowing situations. Baseline conditions, such as lane departure warnings, were incorporated for comparison. To ensure impartiality, the order of test conditions was randomized for each participant. Additionally, post-test questionnaires were administered to gather valuable feedback and insights from the participants' experiences. This section provides a detailed glimpse into the structured evaluation of the Guardian Angel system and its real-world applicability.

Familiarization with the Vehicle: Participants completed a few laps to acquaint themselves with the single-seated Twizy. Specifically, they were asked to take 3 laps, but if requested, they could take more until they felt sufficiently familiarized. They were provided with an explanation of safety measures, including locating the on/off switch inside the vehicle in case of emergencies and the presence of an external operator equipped with a remote emergency stop function, who could stop the vehicle if necessary.

Pre-Tutoring: Participants were shown a 4-min video tutorial showcasing the vehicle's capabilities.

Test Scenario 1: Distracted Driver Scenario

Test 1: Participants engaged in a test of *Scenario 1* to evaluate the system's functionalities. Data from both the vehicle and the driver were logged.

Steps of Scenario 1:

1. **Guardian Angel Activation**: Participants activated the Guardian Angel by pressing a button on the steering wheel. The steering wheel's LED turned green, and a voice message confirmed the activation, with a haptic vibration in the steering wheel as confirmation.
2. **Attentive Drive**: While the driver remained attentive, the system functioned as a lane departure avoidance feature. When the driver approached a lane border, the corresponding LED strip in the A-pillar turned orange to alert the driver. If the driver removed both hands from the steering wheel, the system assumed control and issued warnings.
3. **Distracted Driver**: The LED strip in the windshield bottom turned orange to capture the driver's attention. If the driver became distracted or kept their hands off the steering wheel for too long (3 s), the system escalated, providing additional steering wheel vibrations and flashing lights repeatedly.
4. **System Misuse**: If the system was misused repeatedly, the Guardian Angel would deactivate temporarily and could be reactivated after a few minutes.
5. **System Abuse**: If the system was repeatedly misused, the vehicle was stopped. Initially, voice messages confirmed the system's actions to the driver during these scenarios.

Baseline: The baseline condition involved a lane departure warning. The study did not consider implementing a commercial lane centering system due to the Twizy's lack of such a feature, and it aimed to simulate a realistic sound notification as found in conventional vehicles.

Post-Test Questionnaire: A questionnaire was administered to evaluate the test. Participants rated their experience with each of the GA HMI components using a 7-level Likert scale, which ranged from "poor" to "excellent". This comparison also encompassed the participants' general perception of the GA system as a whole, and the baseline system was included in the assessment. The question to answer was "Overall, how would you rate your experience with the system?" and the 7-level Likert scale for the HMIs had the following wording: "Steering Support", "Lights", "Pre-drive Tutoring", "During Drive Tutoring", "Haptic Icons", "Sounds", "HUD" and "Assembled GA system".

Test Scenario 2: Lane Narrowing Scenario

Test 2: Participants engaged in a test of *Scenario 2* to assess the system's functionalities. Data from both the vehicle and the driver were logged.

Steps of Scenario 2:

1. **Lane Narrowing Entrance**: Upon entering the narrow lane zone, both A-pillar LEDs turned orange, and the lane centering support activated.
2. **Distracted Lane Narrowing Entrance**: If the driver was distracted when entering the narrow lane zone, the corresponding warning initiated more rapidly to ensure driver attentiveness throughout the narrow lane.

For the lane narrowing scenario, the system's baseline was a lane departure warning system.

Post-Test Questionnaire: Participants completed a questionnaire to evaluate the test. The order of Test 1, Baseline 1, Test 2, Baseline 2 was initially presented, but during the actual tests, the order was randomized to ensure no order bias. However, it was decided to maintain the distracted driver scenario before the lane narrowing one, as the latter involved some distractions.

3.6.6 Demo Limitations

Restricted Speed Range: The testing circuit had limited space, allowing speeds no higher than 20 km/h, which restricted the assessment of the system's performance at higher speeds.

Lack of Real Traffic: The circuit lacked real traffic and did not allow for exposure to diverse and real-world driving situations.

Environmental Noises: Due to the open design of the Twizy, external factors like gardeners, nearby vehicles, and even aircraft noises occasionally affected the testing environment.

Operator Guidance: To execute complex scenarios, one test operator was engaged in phone communication to guide the test, while another operator conducted safety checks.

Changing Sunlight Conditions: Testing predominantly occurred during sunny days, which presented challenges for the Head-Up Display (HUD) due to varying sunlight conditions. Given days with high levels of clarity and sunlight, the HUD messages had low visibility.

Limited Elderly Drivers: For safety considerations, only secured drivers were permitted to participate in the tests. This limitation both restricted the number of participants (10) and resulted in a single elderly driver among them (60 years old).

Baseline System Discrepancy: The baseline system simulated a commercial Lane Departure Warning (LDW). However, it should be noted that this LDW system was not an actual commercial system. Furthermore, LDW systems in commercial vehicles are typically designed for highway use and rely on camera-based line detection, whereas the tests were conducted in a small, closed test circuit utilizing a GNSS positioning system. The sounds, however, were taken from an actual LDW of a real vehicle.

Power Limitations: The Driver Monitoring System (DMS), HUD phone, and steering wheel Arduinos could not be powered by the vehicle's battery. Instead, they relied on portable batteries, which limited the duration of continuous system operation to approximately 1 h and 30 min, limiting the continuous duration of each participant's test.

3.7 *Results*

This chapter provides the participants' demographics data and the subjective metrics obtained through post-test questionnaires distributed to the participants, showing participants' user experience with the Guardian Angel system. The questionnaires were administered following each of the two scenarios. Results of the objective metrics are to be published in another paper (Sarabia et al. 2024).

3.7.1 Participants

10 participants took part in this within subject test. One of the goals of the Guardian Angel is to provide a safety system oriented to elderly people, who might have physical restrictions while driving a vehicle. However, to ensure safety of all demonstration participants and surrounding infrastructure, only familiarized drivers who had experience driving the experimental vehicle participated in the demonstrator tests. These drivers were aware of the of the automated functionalities of the vehicle, and were trained to handle any unexpected situations.

Before the tests, a pre-test questionnaire was given to the participants, to obtain the following data: demographic data (age, gender, job), driving experience and driving difficulties (short-sightedness).

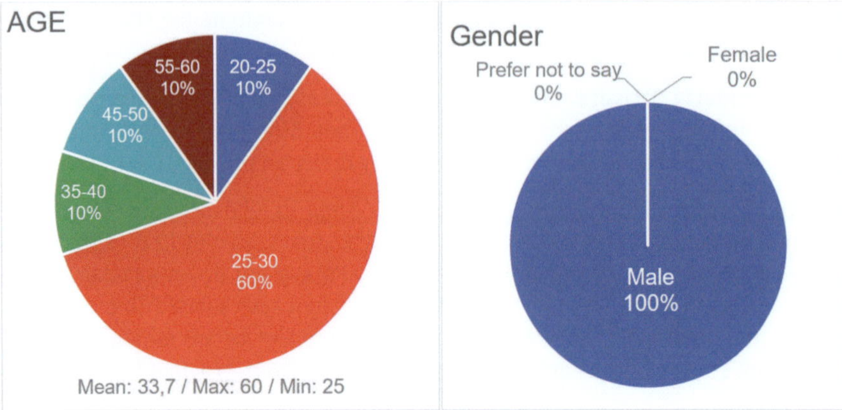

Fig. 14 Age and gender distribution of the participants

Demographic Data of the Participants

All participants in the study were male. This gender composition was not a deliberate choice but rather a reflection of the availability of male drivers during the testing period. Their ages ranged from 25 to 60, with an average age of 33.7, see (Fig. 14).

Driving Experience

The majority of the participants can be categorized as experienced drivers, with a significant portion having more than 2 years of driving experience and a tendency to drive daily, see Fig. 15 through Fig. 18. However, it is noteworthy that they exhibit limited experience with Advanced Driver Assistance Systems (ADAS) usage. Among the ADAS features, Cruise Control (CC) and Lane Departure Warning are most commonly found, yet they are infrequently utilized. Other ADAS features such as Lane Centering or Lane Keeping systems have even lower usage percentages (Figs. 16 and 17).

Participants Limitations:

It's worth noting that 3 participants required corrective glasses for driving, which may affect their field of vision and impacted the accuracy of the Driver Monitoring System's gaze detection.

3.7.2 Distracted Driver Scenario

Participants rated their experience with each of the GA HMI components using a 7-level Likert scale, ranging from 1 (poor) to 7 (excellent), as illustrated in Fig. 19. Each component was compared against the baseline using t-test comparisons. These comparisons also extended to encompass the participants' overall perception of

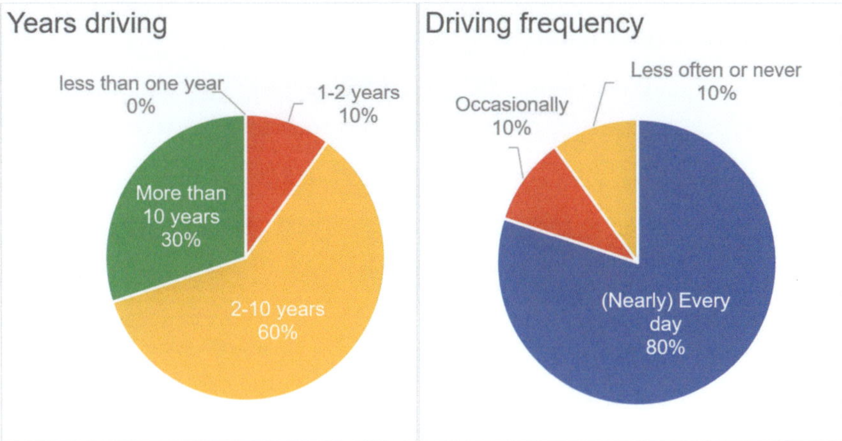

Fig. 15 Driving experience and driving frequency

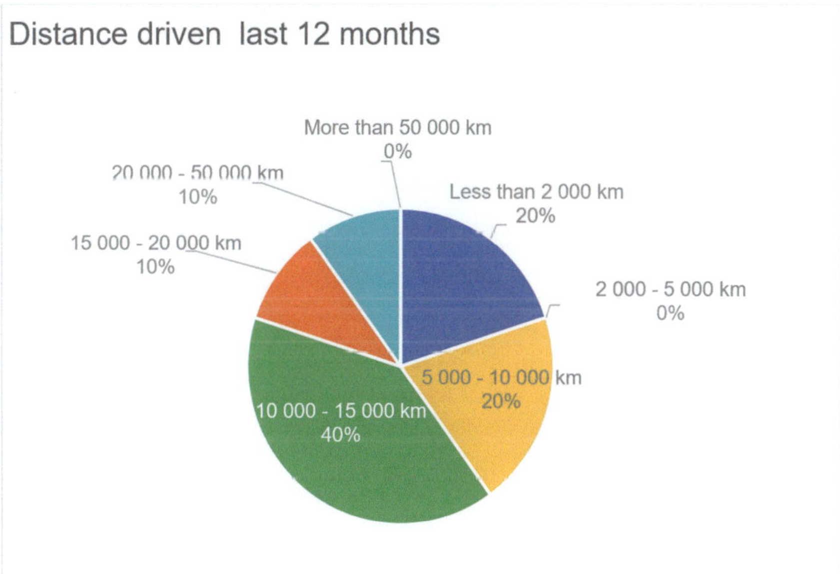

Fig. 16 Driving experience during the last year

the GA system. The p-values shown below were used to determine the statistical significance of the mean differences for each component.

Results indicate that the baseline (BL) system received an average rating of 5.00. In contrast, the Guardian Angel system assembly received a higher average rating of 5.90 though the difference between BL and Guardian Angel did not reach statistical significance ($p = 0.0811$).

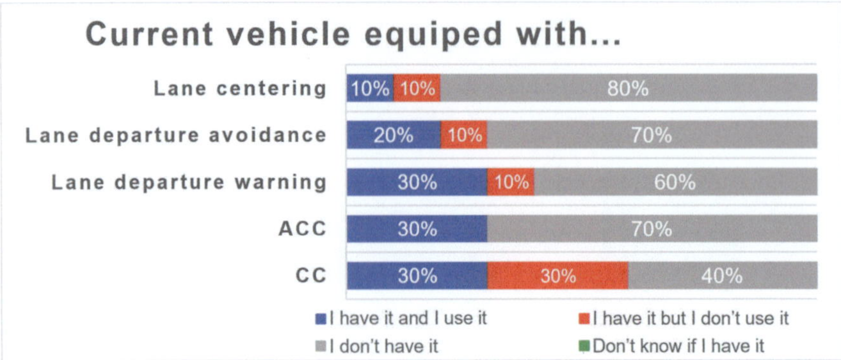

Fig. 17 Experience with ADAS

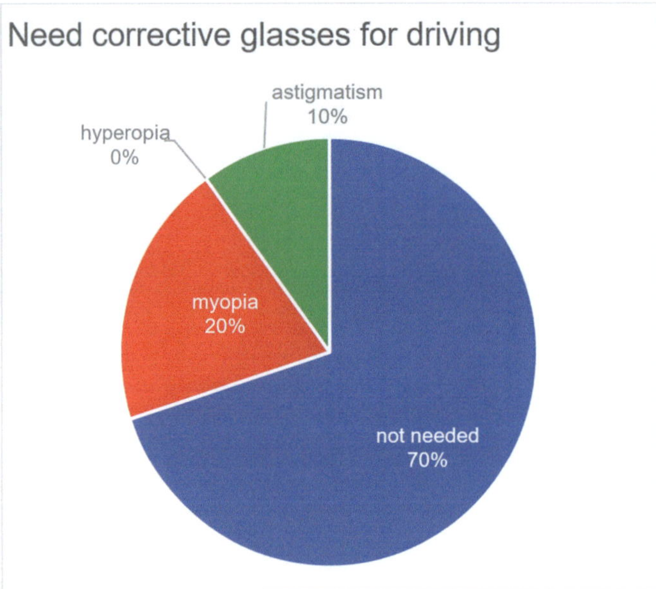

Fig. 18 Ratio of participants using glasses

When examining individual components, the steering support was the most highly regarded, with an average rating of 6.30 ($p = 0.0224$). Pre-drive tutoring also earned a favorable average rating of 6.00 ($p = 0.0319$). Other components, including during-drive tutoring, spatial sounds, ambient lights, haptic icons, and the Head-Up Display (HUD), were individually ranked slightly higher than the baseline, although these differences lacked statistical significance. Consequently, the overall evaluation reaffirms the notion that the Guardian Angel system, with its seven components communicating through visual, auditory, and haptic channels, is perceived as well-integrated

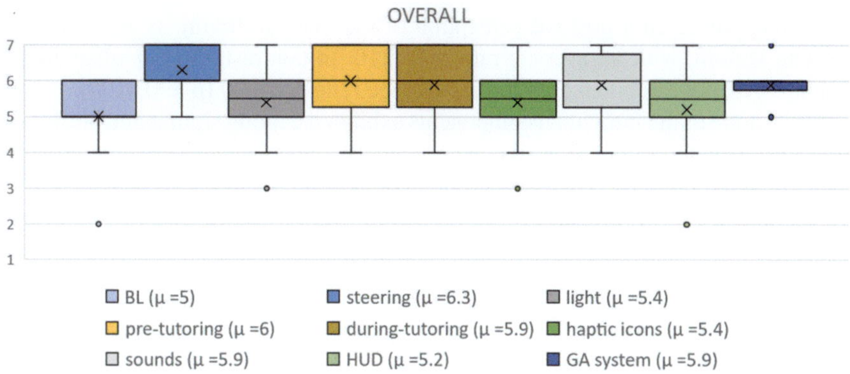

Fig. 19 Results of the participants questionnaires regarding each HMI, the baseline and the GA system for scenario 1, Distracted Driver

and user-friendly. Furthermore, it highlights the system's effectiveness and utility, with the steering support standing out as significantly higher rated in this regard.

3.7.3 Lane Narrowing Scenario

This section shows the subjective metrics acquired through post-test questionnaires provided to the participants during the Lane Narrowing Scenario. Like in the previous one, participants were asked to rate the Guardian Angel HMI components on a 7-level Likert scale too. The results are shown in Fig. 20.

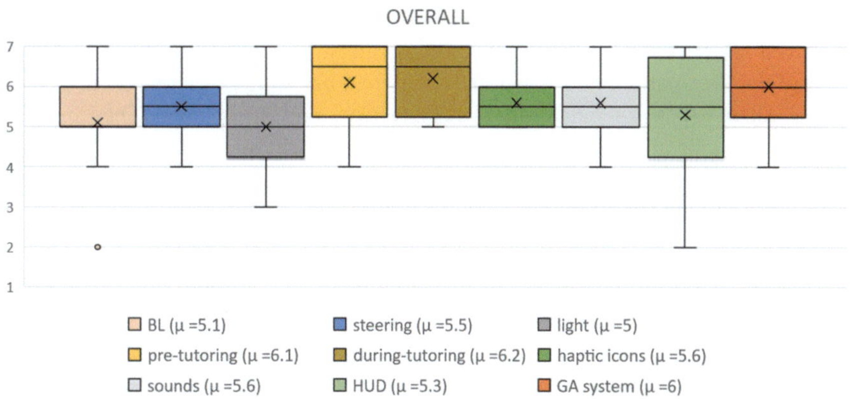

Fig. 20 Results of the participants questionnaires regarding each HMI, the baseline and the GA system for scenario 1, Lane Narrowing

Findings revealed a positive perception of the manual driving system with the baseline system, with an average rating of 5.10. In contrast, the Guardian Angel system assembly earned an even higher average rating of 6.00 ($p = 0.0100$).

While none of the individual components exhibited a rating significantly different from the baseline, all components, except for the lights, achieved average ratings higher than the baseline. The highest-rated components included during-drive ($p = 0.0751$) and pre-drive ($p = 0.0848$) tutoring, followed by haptic icons ($p = 0.3434$), sounds ($p = 0.3434$), and steering support ($p = 0.3732$).

Notably, the Head-Up Display (HUD) and ambient lights received the lowest ratings and displayed the largest variations. This variance might be attributed to the diverse sunlight conditions experienced by participants, as indicated in their post-test debrief comments.

3.8 Demonstration 2 Conclusions

This study presents an evaluation of the Guardian Angel system, offering valuable insights into its performance in different driving scenarios and its impact on user experience. Several key findings emerge from the extensive testing and analysis:

Enhanced User Experience: Participants generally regarded the Guardian Angel system favorably, with a higher average rating compared to the manual driving system with the commercial LDW (baseline). The positive results suggest that the safety benefits are perceived strongly.

Effectiveness of Steering Support: The steering support component stood out as the best-rated element within the Guardian Angel system. Its ability to assist in lane-keeping and correction received significant praise from participants, indicating its perceived effectiveness in enhancing driving safety. Regarding commercial ADAS for lane keeping, participants generally stated that they were not active users of such system. However, actual lane keeping systems have a different scope. They are used in highway applications, at high speed, as a sustained driving support. The Guardian Angel, on the contrary, in this study is used in low-speed scenarios, acting only when necessary and with stronger assistance.

Positive Feedback for Tutorials: Both pre-drive and during-drive tutoring components received favorable ratings, emphasizing their role in educating and guiding the driver within the Guardian Angel system. This educational aspect contributes to a smoother and safer driving experience. Also, there is no significant difference between both scenarios.

Consistent Ratings Across Components: While no individual HMI component exhibited a significantly different rating compared to the baseline, all Guardian Angel components, except for the lights in the lane narrowing scenario, received higher average ratings. This suggests a consistent level of improvement across various aspects of the system.

Ambient Light and HUD Challenges: The ambient lights and Head-Up Display (HUD) components received lower ratings and displayed greater variations among participants. This variance may be attributed to changing sunlight conditions and potential issues related to visibility and usability. For future works, it could be a matter of interest having variable brightness depending on the sunlight conditions.

Limitations of Test Environment: The study revealed certain limitations related to the test environment, including circuit size and speed restrictions, the absence of real traffic, occasional external noises, and power constraints for specific components. These limitations should be considered when interpreting the results.

In conclusion, the Guardian Angel system demonstrates a promising improvement in the driving experience and safety for users, particularly in scenarios involving distracted driving and lane narrowing. The positive feedback from participants, coupled with effective steering support and educational features, underscores the potential of this innovative safety system. However, addressing challenges related to certain HMI components and acknowledging the limitations of the test environment are essential steps toward further refinement and real-world implementation of the Guardian Angel system.

4 Overall Conclusions

Toward the end of this book it remains to conclude what these two vehicle demonstrations have taught us about shaping automated driving to meet societal mobility needs which is the topic of this book. In Chap. 1 of this book we had started with establishing a vision of a larger holistic DAS that creates additional leveraging mechanisms to meet human needs and constraints. Such DAS was conceptualized to include, beside the vehicle, the roadside infrastructure as well as mechanisms to assure the responsibilities and accountability of human drivers for appropriate AD vehicle use. In Chap. 2 we had presented the HADRIAN user-centered approach to shape such a more powerful DAS by identifying the specific mobility needs and constraints of concrete user personas and designed concrete solutions for their use of AD vehicles. In Chaps. 3 and 4 we described specific designs of component innovations for these HADRIAN solutions and the results of evaluating them in driving simulator studies. In chapter 5 finally, we described the conduct and results of two demonstrations where we implemented these HADRIAN component solutions in real vehicles and compared them with two respective baseline systems. In demonstration 1 we evaluated the effects of HADRIAN innovations on safe, comfortable, and acceptable transitions between AD levels for a user who would want to use the AD vehicle to either work or relax with other non-driving related activities during the drive. In this field demonstration we learned that the increased predictability of the AD experience and the adaptive support to drivers to ensure safe and effective take-overs was found to be significantly safer and more comfortable than a baseline system without such innovations. In demonstration 2 we evaluated the effects of an additional safety system, the Guardian Angel which was introduced in Chap. 4 and would become active only if drivers did not appropriately resolve safety critical situations. This was intended to allow drivers with some deterioration of their driving skills, such as elderlies, to continue to use their vehicles in their daily lives. Here we learned that the active steering support along with the tutoring system were the most effective innovations for trip safety and the Guardian Angel functions received high evaluations compared to a baseline system without such innovations.

In both demonstrations, the feasibility of implementing the HADRIAN solutions could be effectively demonstrated and the perceived user benefits was assessed. The demonstration results thereby validate the holistic, user-centered approach of the HADRIAN project. This has, we think, important implications for how the European Union's vision for inclusive mobility can be achieved. In many discourses to this topic, for example at the European Conferences on Connected and Automated Driving,[5] the Transportation Research Arena,[6] or the European Conference on Results from the Road Transportation Research,[7] we see two types of views confronting each other. On one side are those representing public and individual mobility needs, funding research to achieve societal benefits. On the other side, we see the technological research and development organizations who in the end need to sell products that support their business. Both have quite different goals and there is currently no bridge between these two sides as one side cannot directly implement the others demands. Therefore, it remains unclear how the current eco-system of AD developments would be able the shape the kind of technologies that meet real societal mobility needs and also realize the promising high market potential.

Therefore, one paramount implication of the HADRIAN project results is the feasibility and promise of adopting a larger eco-system view within which it becomes possible to create solutions that go beyond the status quo. This starts with a clear vision what problems need to be solved and working with the various stakeholders to collaborate toward a converging solution rather than working on separate component solutions. Such overall system vision can be most convincingly motivated by a human centered system purpose that goes beyond the practical needs of individual component stakeholders. Such overarching system vision is paramount to lead the various stakeholders toward convergence and allow prioritization from a system perspective rather than a mere single stakeholder perspective. Such human-centered system design stands in contrast to today's practice where individual original equipment manufacturers create their AD vehicles with minimal external requirements on the road information infrastructure or responsibilities and seem unable to bring us closer to the hoped-for revolution of the mobility landscapes through AD vehicles. Instead this has brought us smallest common denominator solutions of vehicles driving at SAE L3 under very specific conditions in narrowly defined environments where the usage benefits remain rather small indeed; occasionally somewhat

[5] https://www.connectedautomateddriving.eu/.

[6] https://traconference.eu/

[7] https://research-and-innovation.ec.europa.eu/events/upcoming-events/conference-results-road-transport-research-rtr-2023-02-14_en.

convenient but not sufficient to even touch societal mobility needs or a blossoming market.

Another key lesson for us is that such user-centered approach is needed to formulate development goals that are not otherwise visible from the trajectory of existing technical research and development structures. There, the focus is often on improving what has worked so far and on producing short term solutions. However, a user-centered approach can help to identify the gaps needed to fill toward sustainable market growth. Thereby it is important that a truly user-centered approach often requires a larger holistic view on larger eco-systems to shape the whole system and allocate functions that balance the socio-technical perspectives of human, technology, environment, and organizations (Boy 2013). We see in many research projects across Europe that research can lead to new designs and findings but that these remain often islands of ideas, not implemented in real products on a large scale. On the other side, we do see many new products streaming on the market that are not perceived as useful or that even create societal backlashes as they seem the result of trial-and-error exercises (Endsley 2016). We think that such sunk costs could be reduced or avoided by adopting user centered approaches with a holistic system perspective. However, adopting holistic perspectives takes effort to break out of the increasingly shorter innovation cycles of traditional product developments. Holistic developments may allow in the end for greater returns of investments, but they also require upfront costly investments with amortization timeframes that may go beyond the economic time horizon of many traditional companies. Establishing such a strategic perspective probably requires some central trustworthy agents to not only hold up the banner of human-centered visions of technology developments but also lead toward their implementation and see for their continued operation.

Within the HADRIAN project, we believe that it was possible to establish such human-centered vision and create convergence across many diverse partners and stakeholders toward a larger eco-system of DAS and showed its benefits and outcomes. What has worked on the comparably microscopic scale of a European research project may also work on a larger European scale. With this common vision, we have brought together experts in automated driving research, safety engineers, and human factors to jointly identify solutions, design, and test them in real vehicles. Through this multi-domain collaboration we learned significantly about other domains and were able to formulate a new DAS with the attributes that we wish would exist on a larger scale in Europe.

Concretely, we see following challenges toward user-centered, holistic developments:

First, traditional research and development organizations are structured in the established ways of deeply engrained silos of excellence that follow the traditional lines of educational background and product development processes and organizations. Therefore, there are no experts on holistic DAS but there are experts on vehicle design, human–computer interaction, C-ITS, vehicle and road-side sensors, trajectory planning algorithms, and so on, each searching for solutions within their own domain but virtually nobody leading toward a convergent whole that lies outside of the individual silos. Also, there seems little near-term motivation to adopt such a holistic perspective, rather specific innovations are rewarded such as the numbers of accepted papers, patents, or short-term sales numbers. This naturally leads to many component solutions that work in controlled and possibly contrived test environments as proof of concepts but are often not suitable for real uptake within real-world contexts. Thereby, gaps between the different areas of expertise are considerable and slow down successful integration and collaboration. For example, in the HADRIAN project we noticed large rifts between the world of road infrastructure and automotive development when it comes to formulate the actual communication processes between them; no open coding and decoding tools were available and proprietary solutions had to be purchased and adopted through labor-intensive manual processes. No automation and no open tools were available. Such tools, however, are standard in other domains.

Secondly, there is currently no strong link between societal mobility needs and the design and development of AD vehicles that are currently merely organized around principles of the market. It is currently unclear how building AD vehicles that satisfy societal European mobility needs will be commercially successful for those building them. While research projects can cross such limitations the translation of research outcomes into European reality is not commonly funded and therefore remains well below societal impact.

Finally, there is much work that will need to be done to make holistic, user-centered DAS widely available in Europe. While the development of smart automation is currently focused on the vehicles per se, much more smart automation will be needed for such a large, holistic system. For example, automated categorization and digital formulation of road environment information and its communication to the vehicle will be needed. Also, new curricula, education, training, and responsibility profiles for human drivers will need to be established to not only shape the technology toward the human, but also shape the expectations and skills of drivers toward use of these complex technologies.

We had started this book with the promise of societal benefits of AD vehicles that gave rise to the HADRIAN project. At the end of this book we confirm our belief that high levels of automated driving can have substantial impact on European mobility. In this book we have substantiated that the promised benefits of AD vehicles could be fulfilled by widening our view from a vehicle centered DAS to one that includes the environment and humans. If AD vehicles were in the future not shaped through such holistic, user-centered processes, we fear that they will remain in the realm of slight assistance to human drivers that work sometimes at some places, but in the end are not sufficient to realize societal benefit and long-awaited mobility revolution.

Acknowledgements The planning, preparations, and execution of the field-studies were a joint effort of many partners. Therefore, we want to sincerely thank the following people for their contributions: Karl Lambauer, Kenan Mujkic, Martin Ruidigier, Georg Nestlinger, Kailing Tong, Christian Groß, and Peter Sammer from Virtual Vehicle Research GmbH (Austria); Alexander Mirnig, Magdalena Gärtner, Vivien Wallner, and Jakub Sypniewski from Center for Human-Computer Interaction (Austria); as well as Christos Katrakazas, Marianthi Kallidoni, and George Yannis from NTUA (Greece). HADRIAN has received funding from the European Union's Horizon 2020 research and innovation programme under grant agreement No. 875597. This document reflects only the author's view, the Climate Innovation and Networks Executive Agency (CINEA) is not responsible for any use that may be made of the information it contains. Part of the publication was written at Virtual Vehicle Research GmbH in Graz and partially funded within the COMET K2 Competence Centers for Excellent Technologies from the Austrian Federal Ministry for Climate Action (BMK), the Austrian Federal Ministry for Labour and Economy (BMAW), the Province of Styria (Dept. 12) and the Styrian Business Promotion Agency (SFG). The Austrian Research Promotion Agency (FFG) has been authorised for the programme management.

References

Abdi H (2007) The Bonferonni and Šidák corrections for multiple comparisons. Encycl Meas Stat 19

Bengler K, Zimmermann M, Bortot D, Kienle M, Damböck D (2012) Interaction principles for cooperative human-machine systems. It Inform Technol 54(4):157–164. https://doi.org/10.1524/itit.2012.0680

Boy GA (2013) Orchestrating human-centered design. Springer

C-Roads WG2 (2021) C-ITS infrastructure functions and specifications. 125

Ebinger N, Trösterer S, Neuhuber N, Mörtl P (2023) Conceptualisation and evaluation of adaptive driver tutoring for conditional driving automation. HFES Europe chapter conference Liverpool 2023. https://www.hfes-europe.org/wp-content/uploads/2023/05/Ebinger2023.pdf

Endsley MR (2016) From here to autonomy: lessons learned from human–automation research. Hum Fact 0018720816681350

Eriksson A, Stanton N (2016) Take-over time in highly automated vehicles: Non-critical transitions to and from manual control. Hum Fact. http://eprints.soton.ac.uk/403717/

Eriksson A, Stanton NA (2017) Driving performance after self-regulated control transitions in highly automated vehicles. Hum Fact 16

Marx C, Ebinger N, Santuccio E, Moertl P (2022) Bringing the driver back in-the-loop: usefulness of letting the driver know the duration of an automated drive and its impact on takeover performance. Conference proceedings. Applied Human Factors and Ergonomics Conference, New York

Oviedo-Trespalacios O, Tichon J, Briant O (2021) Is a flick-through enough? A content analysis of advanced driver assistance systems (ADAS) user manuals. PLoS ONE 16(6):e0252688. https://doi.org/10.1371/journal.pone.0252688

SAE International (2021) Taxomony and definitions for terms related to driving automation systems for on-road motor vehicles: (Nr. J3016)

Sarabia J, Marcano M, Diaz S, Zubizarreta A, Perez J (2024) An MPC-based shared control algorithm for hazard mitigation in driver assistance systems: a vehicle study (in production)

Shi E, Bengler K (2022) Non-driving related tasks' effects on takeover and manual driving behavior in a real driving setting: a differentiation approach based on task switching and modality shifting. Accid Anal Prev 178:106844. https://doi.org/10.1016/j.aap.2022.106844

MIX
Papier aus verantwortungsvollen Quellen
Paper from responsible sources
FSC® C105338

If you have any concerns about our products,
you can contact us on
ProductSafety@springernature.com

In case Publisher is established outside the EU,
the EU authorized representative is:
Springer Nature Customer Service Center GmbH
Europaplatz 3, 69115 Heidelberg, Germany

Printed by Libri Plureos GmbH
in Hamburg, Germany